TALES OF
Remarkable
BIRDS

TALES OF
Remarkable
BIRDS

DOMINIC COUZENS

B L O O M S B U R Y
LONDON · NEW DELHI · NEW YORK · SYDNEY

Bloomsbury Natural History

An imprint of Bloomsbury Publishing Plc

50 Bedford Square

London

WC1B 3DP

UK

1385 Broadway

New York

NY 10018

USA

www.bloomsbury.com

Bloomsbury Natural History and the Bloomsbury logo
are trademarks of Bloomsbury Publishing Plc

First published 2015

British Library Cataloguing-in-Publication Data

A catalogue record for this book is available from the British Library.

ISBN: HB: 978-1408-1-9023-4
ePDF: 978-1-4081-9024-1
ePub: 978-1408-19025-8
10 9 8 7 6 5 4 3 2 1

Designed by Nicola Liddiard, Nimbus Design

Printed in China

To find out more about our authors and books visit www.bloomsbury.com
Here you will find extracts, author interviews, details of forthcoming events
and the option to sign up for our newsletters.

Photos: p.1 Superb Fairywren; pp.2-3 Emperor Penguin chicks huddling;
p.5 Northern Double-collared Sunbird

Contents

Introduction 8

EUROPE 14

Northern Wren: Roost site available, not everyone need apply **16**

Great Spotted Cuckoo: A cuckoo, yes – but not as we know it **20**

Great Grey Shrike: Catching on to a neat idea **26**

Common Crossbill: A bird with leftist tendencies **32**

Eurasian Oystercatcher: The oystercatcher's trade unions **36**

AFRICA 40

Sunbirds: Hovering might be catching **42**

Ostriches: The benefits of sharing a nest **46**

Straw-tailed Whydah and Purple Grenadier:
     The strange case of the avian stalker **52**

Boubous: It takes two **56**

Widowbirds: A tale of two tails **60**

ASIA 66

Greater Racket-tailed Drongo: The life of a professional agitator **68**

Yellow-browed Warbler: The wrong-way migrant **72**

Pheasant-tailed Jacana: Children of the lily-pads **76**

Arabian Babbler: Keeping its friends close... **80**

Swifts and swiftlets: Living in the dark **84**

AUSTRALASIA 88

White-winged Chough: Our family group needs some extra help **90**

Fairywrens: What is the significance of the flower gift? **94**

Great Bowerbird: Stage managing a nuptial bower **98**

Southern Cassowary: Don't mess with this big bird **104**

Varied Sittellas: The benefits of working together **110**

NORTH AMERICA 114

White-throated Sparrow: A tale of two sparrows **116**

Black-capped Chickadee: Memories of garden birds **120**

Cliff Swallow: Unnatural selection **124**

Harris's Hawk: The hunter-gatherer **128**

Marbled Murrelet: Breeding in a different world **132**

SOUTH AMERICA 136

Andean Cock-of-the-Rock: Working together with its friends **138**

Toucans: Why a big bill pays **144**

Antbirds: Following the ants **150**

Tanagers: The crown jewels **154**

Hummingbirds: When the humming stops **158**

ANTARCTICA 164

Rockhopper Penguin: The most unloved egg **166**

Albatrosses: Masters of the oceans **172**

Emperor and Galapagos Penguins: A tale of two penguins **178**

Sheathbills: The basement cleaners **184**

Wandering Albatross: A slow dance to success **188**

ISLANDS 194

Swallow-tailed Gull: Making the most of dark nights **196**

Megapodes: The patter of great, big feet **200**

New Caledonian Crow: The world's cleverest bird? **206**

Blue Bird-of-paradise:

   Figs and fruits turn paradise upside down **210**

Extinctions: Islands: lands of lost birds **216**

Further reading **220**

Index **221**

Acknowledgments **224**

INTRODUCTION

This book is a celebration of bird behaviour around the world. It is a small taster for a great feast. Birds do some extraordinary things, and to cover all their solutions to the trials of life would take many volumes. A taster is intended to whet the appetite, so the function of this book is really to entice the reader to find out more about what birds get up to around the world, by reading further books, trawling the internet or going into the field to watch. The next discoveries, after all, are often in the backyard.

How does one go about selecting some stories to reflect the diversity and complexity of avian lives? I have used three main criteria: a worldwide spread of stories, a spread of behaviours from across the spectrum of what birds do (migrating, feeding, incubating eggs, and so on), and my own personal preferences. I have tried to avoid what we in Britain call 'old chestnuts', stories that most people have already heard and will not come to fresh. As a result, some of those in this book are quite obscure, and I make no excuses for that.

This book is divided into eight sections to keep a wide geographic breadth of stories. Most sections cover genuinely biogeographical entities: North America (Nearctic), South America (Neotropical, which includes Central America), Africa south of the Sahara (Afrotropical), Australasia and the Antarctic. However, for convenience Europe is treated as an entity because of the high level of research traffic there, and one story from the Asian part of the Palearctic is included in the Asian section. Furthermore, the world's Islands are treated in a section on their own.

Although the spread of stories largely reflects the wide spectrum of different bird behaviours, I have tried to cover themes that are relevant to the region, where this is possible. For example, the obligate following of army ant swarms is best developed in the Neotropics, and in Australia there is an unusual high percentage of group living birds. You might also argue that duetting is especially well developed in Africa, and that incubating an egg independently of a bird's skin is virtually confined to Australasia and Oceania. Nevertheless, few behaviours have real geographical limits.

Opposite: A Grey-headed Albatross protects its chick against the threat of a skua overhead.

While a book on global bird behaviour needs a suitably complete geographic reach, it also needs to cover the main ornithological bases in regard to the different types of behaviour. It will never get near to complete, since within every division of bird biology (e.g. breeding), there are numerous subdivisions and subdivisions of subdivisions. Indeed, the sheer scope of behavioural research is simply overwhelming – where do you start? In preparing this book I divided a bird's life into its various compartments and made sure that, across the work, as many as possible were covered. Hopefully readers will find that their pet subject is included somewhere.

To give you an idea of the range of subjects covered, here is something of an alternative index to them, with the type of behaviour and the species or families concerned:

Above: A male Ostrich calling and displaying.

Roosting: Northern Wren (Europe), Varied Sittella (Australasia), Hummingbirds (South America)

Incubation: Rockhopper Penguin (Antarctic), Pheasant-tailed Jacana (Asia), Ostrich (Africa), Micronesian Scrubfowl (Island)

Nest-site: Marbled Murrelet (North America) • **Infanticide:** Pheasant-tailed Jacana (Asia)

Group-living: White-winged Chough (Australasia), Arabian Babbler (Asia)

Vocalisations: Boubous/Gonolek (Africa), White-crowned Sparrow (North America)

Duetting: Boubous/Gonolek (Africa)

Pair-bonds: Fairywrens (Australasia), Great Grey Shrike (Europe), Albatrosses (Antarctic)

Sexual selection: Long-tailed Widowbird (Africa)

Display: Great Bowerbird (Australasia), Blue Bird-of-paradise (Islands)

Lek: Andean Cock-of-the-Rock (South America)

Brood parasitism: Great Spotted Cuckoo (Europe), Whydah/Grenadier (Africa)

Parental care: Pheasant-tailed Jacana (Asia), Rockhopper Penguin (Antarctic), Marbled Murrelet (North America)

Commensal feeding: Greater Racket-tailed Drongo (Asia)

Food-storing: Great Grey Shrike (Europe), Black-capped Chickadee (North America)

Resource partition: Oystercatcher (Europe), Tanagers (South America)

Communal foraging: Harris's Hawk (North America)

Optimal foraging: Swallow-tailed Gull (Islands)

Ant-following: Antbirds (South America)

Scavenging: Sheathbills (Antarctic)

Nest-robbing: Toucans (South America)

Mobbing: Greater Racket-tailed Drongo (Asia)

Evolution in action: Sunbirds (Africa), Cliff Swallows (North America)

Comparative ecology: Tanagers (South America), Penguins (Antarctic)

Memory: Black-capped Chickadee (North America)

Intelligence: New Caledonian Crow (Islands)

Echolocation: Swiftlets (Asia)

Flight style: Hummingbirds (South America), Albatrosses (Antarctic)

Footed-ness: Crossbill (Europe)

Physical intimidation: Toucans (South America)

Migration: Yellow-browed Warbler (Asia), Albatrosses (Antarctic)

Birds and people: Southern Cassowary (Australia), Cliff Swallow (North America)

Conservation: Hawaiian Honeycreepers (Islands)

Broad though this list seems, it can only ever scratch the surface of the complexity of bird biology. It would not be an exaggeration to suggest that almost every bird in the world has the capacity to amaze and surprise scientists. It just depends what species and aspects are chosen for study. You could write a book with exactly the same title and premise, and choose 40 completely different stories to the ones chosen here.

There is a very strong bias of personal preference in the book, and it is relevant to explain this. I have included species that I have seen in the wild. I might not have seen the specific behaviour, but seeing a Malachite Sunbird, for example, gives you some insight into what it might look like when hovering. While this is not a scientific way of selecting a story, it is a personal celebration of ornithological science.

Finally, there is a slight bias in the selection towards good-looking species. I make no apology for this. Great Snipes and Andean Cock-of-the-rocks both display on a lek, so which one do you choose – the brown, cryptically coloured one or the brilliant crimson and black one? Books, ultimately, are works of art as well as agents of communication. So, for every Arabian Babbler there has to be a Blue bird-of-paradise, and much as a Sooty Shearwater flies over the ocean like a dream, a Light-mantled Sooty Albatross does the same while bursting with star quality. In the end, the book stands or falls on its selection of stories. I had to leave many great ones out. But then, that is the glory of bird behaviour; the more you discover for yourself, the more you want to find out. This informs personal discovery, and drives the scientific process.

In many ways, this book is a tribute to all the scientists and field workers, conservationists and authors, whether they are professionals or amateurs, who go out there and discover new things about birds and other wildlife. They are a powerful army of independent minds, melding enquiry with passion. This work is a very small reflection of what they do. They are among the heroes and heroines of our age.

Finally, it is my fond hope that this book will further the reader's passion for ornithology. Goodness knows, birds and other wildlife around the world need our help at the moment. I write this just as a report by the London Zoological Society says that global wildlife populations have halved in the last 40 years. If books such as this can somehow help inflame a person's passion so that they are further inspired to fight against this trend, then they will be worth the effort.

Dominic Couzens, Dorset, UK, October 2014

Opposite: The feisty Greater Racket-tailed Drongo.

●

EUROPE

NORTHERN WREN

Roost site available, not everyone need apply

You don't need to have many encounters with the Northern Wren (*Troglodytes troglodytes*) to appreciate two things about it: it is very small, and it is highly strung. Its reduced size is an adaptation to a life of probing into very small crevices and secret passageways deep and low in tangled vegetation, where few other birds will go. This highly strung nature plays out in the Northern Wren's incessant noisiness; whenever you enter into wren habitat you will hear it before you see it, either giving an incessant, testy 'teck' call when it is disturbed, or producing a slightly over-elaborate song, in which more than 100 separate notes are squeezed into seven seconds, with the effect of a sports commentator describing the end of a 100-metre sprint. The wren's loud-mouthed nature isn't a sham. The songs and calls are merely the outworking of an aggressive and excitable temperament, caused in part by the wren's need to maintain a territory all year round. These birds must maintain their boundaries to feed and breed successfully, leading to inevitable conflict and rumpus. Skirmishes are frequent and quite often physically violent.

The Northern Wren's combination of small size and combustible nature can on occasion lead it directly into a perfect storm of trouble. This happens particularly on chilly winter nights, when the temperature drops towards freezing, or there is particularly high wind or heavy rain. On such occasions it is a disadvantage to be small. A little body confers a relatively large surface area in relation to overall volume, meaning that heat is lost more rapidly than it is for larger birds. Heat loss on cold nights can easily be fatal, since any fat reserves built up during the day simply burn up before they can be replenished. Wren populations often crash during hard northern winters.

However, there is a way in which wrens can ameliorate this heat loss, and that is to find a colleague to make a night time huddle. By making physical contact, two or more birds effectively make themselves into a larger organism with a more favourable surface-to-volume ratio, which can make the difference between life and death. Many species around the world do this.

There is, though, one problem for the Northern Wren, and that is its volatile temperament. It is truly a loner, occupying an exclusive territory that involves being aggressive to all its neighbours. You can imagine that, if you have had a violent skirmish with a peer in the afternoon, you'll hardly want to find yourself needing to snuggle up with your combatant a few hours later. Not surprisingly, Northern Wrens generally don't cuddle unless they have

to. They are adept at squeezing into crevices in rocks and walls and dense vegetation, and on many a night they are fine.

But during freezing spells the wrens have to swallow their pride, and gather a few to a hole. There are records of mass roosting involving tens of birds – for example, a single nest box that contained 61 individuals, and a small piece of thatched roof that hosted 30 together. It is difficult to imagine that such cramped conditions would be pleasant for any bird, but for the anti-social Northern Wren it must be particularly stressful. When huddled in a micro-space, the birds face inwards towards the middle, so their wings and tails face outwards, and they pile one on top of the other to make a series of layers of small birds. How desperate they must be.

Studies on the particular sites used for communal roosting have shown that, rather than birds dropping in from outside to the nearest possible shared crevice, sites used for gatherings are traditional, and used from year to year. This begs an interesting question: are sites known in advance, part of local wren culture, passed on from generation to generation? Or do the birds simply recognise a good cubbyhole for what it is? The latter seems unlikely, because some birds are known to commute as far as 2km to get to the right spot, and that would take them well beyond their territorial boundaries and perhaps to an unknown place.

Above: The Northern Wren is vehemently territorial. In the early morning, a male has been known to sing 200 times in an hour.

In fact, the act of gathering for a mass roost is quite an event in itself. The owner of the territory in which the 'hotel' is found is the one that initiates the assembly, making loud bursts of song and apparently flying around its patch, giving every impression that it is advertising for roost-mates. Presumably the signal passes around the neighbourhood and, little by little, birds find their way to the entrance. Once again, it seems strange that individual birds that might well have been at each other's throats earlier in the day or earlier in the season, are now contemplating being bedfellows – and by invitation, too.

However, what actually happens is slightly different. In the preceding paragraphs you might have expected that birds attending the nightly gathering would simply arrive at the appointed crevice, take their place indoors and cope with the night, with a truce breaking out, however uneasy. That impression is not quite right. It seems that in real life, some individuals are refused entry. As far as can be understood from the evidence, females are invariably allowed to come in. But some male wrens have to fight their way in, literally, or are evicted.

The exact nature of the evictions isn't entirely clear. Sometimes groups of males fight physically outside the roost entrance, at other times persistent birds force their way in, and on other occasions they are turned away and return, presumably, to their own roost sites. In the latter case, one can only imagine that their survival chances are impaired.

While we can speculate exactly what might be going on, one scenario is particularly interesting. Imagine you were the owner of the territory and acted as the summons to the roost. Suddenly, you are confronted by a neighbour who desperately needs to use your roost site. You may have spent months in conflict with this irritant, a threat to your territory and your chances of breeding. Turning them away could be of considerable benefit, well worth an evening fight and a little extra energy expended.

In such a case, being small, as well as highly strung, could suddenly seem advantageous.

Opposite: The winter is a tough time for birds as small as the Northern Wren, just 9–10cm long from bill to tail. Seasonal mortality is very high.

GREAT SPOTTED CUCKOO
A cuckoo, yes – but not as we know it

It was Pliny the Elder in the 1st century AD, who first recorded something odd about the Common Cuckoo (*Cuculus canorus*). Observing smaller birds, now known to be egg-hosts, feeding outsized cuckoo fledglings by almost putting their heads in the parasites' mouths, he denounced the cuckoo for eating the parents that had put their efforts into bringing it up. Ever since then the Common Cuckoo has been viewed as something of a rogue, or worse. Down the centuries people have been genuinely horrified by its lifestyle, sub-contracting the care of its young to smaller hosts, and 'repaying' the host by wiping out its own brood. The cuckoo has at best a questionable reputation.

In more recent times we have expanded our knowledge, and now we know that the cuckoo family contains some 140 species. It is a measure of the Common Cuckoo's infamy that people are still surprised to hear that a small majority of cuckoo species actually build nests and bring up their own young. Yet this is the case: there are domesticated cuckoos, such as North America's famed Roadrunner (*Geococcyx californianus*), and parasitic cuckoos, like the Common Cuckoo – 'good' Cuckoos and 'bad' Cuckoos, if you like.

It so happens that in Europe, where the Common Cuckoo is so well known, there are actually two species of cuckoo. The 'other' species, the Great Spotted Cuckoo (*Clamator glandarius*), almost certainly occurred under Pliny's nose, as it breeds in Italy where the great thinker was born. However, it was not mentioned by him and has generally slipped under the radar, in no way attracting the folklore, wonder and sometimes disgust associated with its celebrity relative. The Great Spotted Cuckoo, though, is not a bird to overlook in life: it is large, noisy and strikingly patterned, with rows of neat white dots on its grey upperparts, and a peachy wash to its neck and upper breast. It also has a ragged crest and a long tail that it raises often. You cannot miss it. But evidently, as far as its behaviour is concerned, it has largely escaped notice.

The question is is this: is there a 'good' cuckoo in the heart of Europe?

At first glance, the signs are hopeful. For one thing, Great Spotted Cuckoos do not parasitise smaller birds, those unfortunate sorts such as warblers and Robins (*Erithacus rubecula*) that always seem so feeble and helpless when faced with Common Cuckoos. However, Great Spotted Cuckoos are obligate brood parasites. Their hosts are crows and magpies (*Pica pica*), birds of similar size to themselves – a fact that has a significant effect on their behaviour. In contrast to Common Cuckoos, where the female is able

to act alone when it steals up to a vacant host nest and lays its egg, pairs of Great Spotted Cuckoos must act as a team. A magpie or a crow could cause a serious injury to a nest-visiting Great Spotted Cuckoo, especially if the latter was trapped inside the magpie's domed nest. So the male Great Spotted makes itself as conspicuous as possible in the vicinity of the chosen nest, calling and perching in full view to distract the potential hosts, while the female takes advantage of the diversion to dash in and out as quickly as possible, often depositing her egg within 10 seconds.

Another difference between the two cuckoos is that the Great Spotted nestlings do not dump the host egg or chicks out of the nest: that behaviour that so shocked people when it was first described by Edward Jenner (the same man who instigated the first vaccination in 1788). And indeed, neither does the female remove an egg when she visits, as the Common Cuckoo does. This is partly an adaptation to prevent the accidental removal of another cuckoo egg or chick because – again unlike the Common Cuckoo – a single female Great Spotted Cuckoo may lay several of her eggs in the host nest. But

Above: *When pairs of Great Spotted Cuckoos raid nests they work as a team. The male perches conspicuously out in the open and calls noisily.*

it would also point towards a much less ruthless strategy than that adopted by the better-known cuckoo species.

You might think that, so far, the Great Spotted Cuckoo is fast solidifying a place in the 'good' cuckoo camp. And when you hear that, on occasion, pairs of Magpies or crows have been known to raise young of their own alongside a Great Spotted Cuckoo in the same nest, you might conclude that this is a case of brood parasitism 'lite', an acceptable version of an ugly strategy. However, you would be quite mistaken.

The Great Spotted Cuckoo is, after all, a parasite, and it adopts a number of strategies that are every bit as merciless as those of the Common Cuckoo – let's just say that they are sneakier.

For one thing, the visit of the female Great Spotted is rather more destructive to the host than might at first appear. True, the laying female doesn't take a host egg away and eat it; but that doesn't mean that it leaves the host clutch alone. Researches examining violated nests often find that some of the host's eggs are chipped or cracked, whereas very few nests that

*Opposite: A recently fledged Great Spotted Cuckoo perched by its foster parent, a Hooded Crow (*Corvus cornix*). Juveniles are distinguished from the adult Cuckoos by the chestnut in the wing. **Above:** It is a common strategy among brood parasites for their chicks to eject the eggs and chicks of the host. Here a newborn Common Cuckoo chick is trying to remove a newborn Reed Warbler (*Cercotrichas galactotes*) from the nest.*

are not parasitised have such problems. It is clear that sometimes the female Great Spotted Cuckoo deliberately damages the host eggs, either by pecking at them or, more deviously, dropping her own harder-shelled egg upon them as she lays it. Either way, she typically makes some attempt to disable her own chick's opposition.

The undercover tactics continue into the nestling stage. Remember, the newly hatched Great Spotted Cuckoo does not actually destroy any of its host's eggs or chicks directly, but it will try to out-compete them to death. The cuckoo chick usually hatches first, sometimes even when the host egg has a head start. Cuckoo eggs often begin development in the mother's oviduct, before incubation, and in the case of the Great Spotted Cuckoo they hatch in only 12–15 days, as opposed to 20 in the case of a magpie. To compound its advantage, the cuckoo chick also grows much faster than magpie or crow chicks, fledging in double quick time, sometimes in as little as two weeks. Meanwhile the laggard magpie or crow chicks leave the nest between 21 and 30 days after hatching. The speed of development is highly significant, for two reasons. Firstly, the cuckoo chick will be able to commandeer a lion's share of any food that is provided by the parent, simply by being more active than its peers and perhaps intercepting food meant for them. Secondly, it can actually physically assault the other offspring in the nest, especially as it grows older. It won't kill them, but could impair their attempts at begging; this is usually enough to remove the opposition.

A final adaptation is still more remarkable. It seems that the young cuckoo actually mimics the sound of the host's chicks in the nest, be they crows or magpies. And not just that – it mimics the sounds that magpie chicks, at least, make when they are particularly hungry and desperate for food. magpie parents are known to respond to differing hunger calls in the nest, responding to the most earnestly beseeching ones. The Great Spotted Cuckoo chick ensures that it gets fed by always sounding hungry, even when it is in fact satisfied, or even satiated. The idea, therefore, is not just to get fed, but to prevent its rival chicks from being fed. Not surprisingly, the genuinely hungry host chicks, way behind the cuckoo in size and influence, are sometimes fatally neglected.

So is there a 'good' cuckoo in Europe? From the above evidence, we can reasonably conclude that there isn't after all. Not all cuckoos can change their spots.

Above: Wherever they go parasitic cuckoos are mobbed by
small birds, presumably aware of their threat. In this case a
Corn Bunting (Miliaria calandra) gives a Great Spotted
Cuckoo grief, despite never being one of its hosts.

GREAT GREY SHRIKE
Catching on to a neat idea

This is the story about how a bird tried something out, and it worked
so well that, in time, the new trick became the norm among the bird's
population. The trick spawned a cascade of related behaviours, and
eventually became foundational to the bird's lifestyle.

Shrikes the world over are famous for their unique habit of impaling prey
on sharp points – typically thorns on bushes, but also, on spikes of barbed
wire and other human artefacts. In the breeding season captured prey is
stored in so-called 'larders', a bizarre and unnerving field sign of a shrike's
presence. It is genuinely creepy to come across the impaled bodies of rodents,
insects and the occasional small bird, all grimly adorning the same branch
or series of low branches. It is reminiscent of the human practice, sometimes
carried out in times of war, of hanging the mutilated corpses of enemies from
trees, as an example of the terrible fate that may await them.

Nobody was observing shrikes back in the distant past when they
first developed their impaling habit, but it doesn't take a large leap of the
imagination to infer how it happened. Shrikes are very unusual in being
songbirds that have adopted the lifestyle of a bird of prey, such as a hawk or
falcon. They therefore tackle quite large prey items, such as small mammals,
lizards and birds, as well as heavy-bodied beetles and dragonflies. The
victims are dispatched by a bite to the back of the neck and subdued by
the shrike's large claws. The problem is that, once killed, the prey has to be
handled. It is usually too risky to leave the body on the ground where it lies,
because it makes the predator vulnerable. The hard-won meal must be taken
somewhere, preferably above ground. It seems unlikely that the shrike made
the leap straightaway to impaling the corpses; instead, one can imagine a
three-stage process.

The first would simply be to carry the corpse up to a perch, whereupon the
bird could hold it down with its claws. After this happened countless times,
the next logical step would be to wedge the body in some kind of cleft, such
as the fork of a branch, to prevent it falling and making it easier to strip flesh
from it. A number of avian predators other than shrikes wedge prey, and
even among the shrikes, some individuals practise this more than actually
impaling (there is evidence that they follow techniques learned early in life,
and may not progress to impaling at all). However, using thorns as hooks and
hanging the prey from them is a step up altogether. A body can be lodged
securely, ensuring that it stays in place, and the predator can dismember it

Above: *Shrikes are unusual for being highly predatory songbirds, taking vertebrate as well as invertebrate food. They are also boldly marked, a form of passive territorial advertisement.*

little by little, at leisure. One day, perhaps by accident, a shrike used a hook in this way.

Once the practice of impaling (and indeed wedging) caught on, there would be an immediate consequence – food could be stored. On a good day for hunting, excess need not be wasted, but could be cached. Straight away, hunting efforts could be modified for the better, and provisions kept for times of greater need.

And from that very moment, caching became pivotal in the behaviour of some shrikes, a good example being the Great Grey Shrike (*Lanius excubitor*).

*Above: A European Robin (*Erithacus rubecula*) falls foul of a Great Grey Shrike. Birds are a particularly important food item in the winter.*

For example, just as soon as the process of caching became widespread, it could also become competitive. Some birds would be better hunters and their caches would appear larger and more impressive. One male might show off its impaling prowess to another bird and use it as a territorial marker. Most bird species defend their territories exclusively by song, by display and eventually by physical combat, but don't leave anything visible. But how much better is a physical, unarguable show of superiority: more items in a cache or, better still, larger items? It takes physical strength to catch a mouse. Why not display the mouse to show off your prowess, just as human hunters seem compelled to hang up skulls and make trophies to prove theirs?

Researchers in Poland mapping Great Grey Shrike caches have found strong evidence that the birds do indeed use impaled bodies as flags of their superiority. For example, the birds don't create larders all year round, but have seasonal peaks and troughs. One such peak occurs in the shrikes' pre-breeding stage, when the need to define territories is at its keenest, suggesting a link between larders and territory. The caches also tended to be found on the boundaries between individual territories, and usually in the open in prominent locations; many mammals do exactly the same when they scent-mark their territories and leave faeces or feeding signs. Even more telling, during this same pre-breeding stage most of the displayed food remained uneaten. If the shrike was not going to consume the cache, this is strongly suggestive that it was put there for another purpose: bravado.

Furthermore, while caching would seem to have an effect on rivalries between males, it also affects the relationship between male and female. One rather obvious effect is that a male can provision his mate. During the breeding season, Great Grey Shrikes subtly move the location of their larders away from the edges of the territory, to sites close to the nest. This is highly convenient for the female, the partner responsible for all the incubation of the eggs and for looking after the chicks in their early stages. If there is a fulsome cache of food near the nest, the female can simply take from it as and when she needs to. In this way the cache quite literally acts as a larder.

It is hardly surprising that the rather intimate act of providing food for a mate plays a part in bringing the sexes together in the first place, and also then goes on to spin a tangled web in shrike-to-shrike relationships. In a social system where males show off by putting their larders on public display, it is inevitable that the females will take notice. Just as males can intimidate

other males by hanging up large, difficult to subdue prey items, so they can impress potential partners in the same way. It is hard to argue that a big cache is a sign of male quality. It is a no-brainer: females preferentially select good hunters.

Of course, not every female can pair up with a good quality male. Great Grey Shrikes are socially monogamous, teaming up with one individual of the opposite sex in order to make a breeding attempt. The pair must work together in order to bring up young successfully. But this does not necessarily mean that either partner is 'faithful' in the genetic sense: extra-pair liaisons are frequent. For a male partner, it makes sense to copulate with as many females as possible, while for a female partner, an extra-pair liaison with a top quality male may serve to offset any failings in her social mate.

And how are these extra-pair liaisons brokered? Using caches, of course. It has been found that males offer gifts from their larders in order to appeal to a potential extra-pair co-conspirator. And furthermore, they cherry-pick the best gifts in order to pay for sex. While they typically provide 16 per cent of their mate's daily food requirements, these special gifts may comprise a whopping 66 per cent of a potential extra-pair female's needs. Not surprisingly, they are often successful.

Who would have thought that a bird's invention of a new way to handle food would lead to developments like this?

Opposite: A Striped Field Mouse (Apodemus agrarius) *is a high-value offering to give to a mate – probably best for the 'mistress'.*

COMMON CROSSBILL

A bird with leftist tendencies

Many a birder in Europe and North America is familiar with the Common Crossbill (*Loxia curvirostra*). This handsome, chunky seed-eater is famous for having crossed mandibles: the lower grows upwards and to the side of the upper mandible, eventually sticking out visibly above it. This unique feature confers a degree of well-earned celebrity upon the Common Crossbill and its immediate close relatives. Seeing a crossbill is always something of a treat, simply because it is so very 'different' to other birds.

If you see a crossbill in the wild, make sure you get as close to it as possible. If you do, you might just notice a rare example among birds of what is known as 'footedness', which is the approximate equivalent of left- or right-handedness in human society. Individual crossbills handle the cones from which they extract seeds either with their right foot or their left foot. You can easily tell which orientation a bird has if it snips off a cone and takes it to a perch, because it will then stand with one foot gripping the perch and the other, its favoured foot, holding the cone. A bird holding a cone in its right foot will have its head and bill to the left of the cone from its point of view, and vice versa. Individuals consistently use one foot in preference the other, just as most of us feel more comfortable using one arm or foot as opposed to the other. They rarely if ever use their 'wrong' foot.

Among crossbills, however, the preference has an obvious physical connection: the way the bill itself crosses. This is where things get really interesting. When a nestling crossbill hatches, its bill is at first straight, with the mandibles fitting neatly together. In the early days the young bird receives food from its parents in same way that every other bird chick does, but in contrast to many, it remains in this dependent state for some considerable time. It isn't until its 27th day of life, no less, that the lower mandible begins to twist in to one side or the other of the upper mandible, and not until the 38th day can the youngster even start to work on the cones on which it will subsist for the rest of its life. After 45 days, more than six weeks, it extracts seeds from the cones efficiently enough to fend for itself. If proof were needed that the bill had to cross in order for this unusual bird to survive, then this is it.

Opposite: The unique bill of the Common Crossbill allows it to reach seeds in cones that are still ripening. The bird's strong legs give it great agility in the treetops.

The divide among crossbills is even more profound than it sounds. Not only does the mandible physically twist either left or right, but the jaw muscles controlling the mandibles are larger on the side to which the bill is deflected, adding to the asymmetry. This almost certainly means that, once the twist to either side has occurred, the individual will be oriented in this way for life.

Most unusually, and in contrast to humankind's hand and foot orientation, the proportion of left-billed to right-billed individuals is approximately equal, and as far as is known, is the same for both sexes. This suggests that there are no survival implications, and presumably no disadvantages either way. There is also no suggestion that left-billed crossbills preferentially pair up with a mate of the same, or different orientation.

However, the difference in practice is a major one. The crossbill's orientation is central to its feeding method. If you've ever wondered why the bill is crossed at all, the answer lies in the fact that, as the bill is shut the lower mandible moves laterally in relation to the upper one. If the bill tip is placed at the tip of a pine cone scale, the act of shutting the bill will force the scale open so that the bird can scoop the seed out with its tongue. The force of closing the bill is very strong, meaning that crossbills can reach seeds that are still firmly wedged between the scales and not yet ripe, but in the process of ripening. Birds not blessed with the crossbill's trick simply cannot tap into this rich resource, leaving the field free for these highly specialised birds.

Why, though, should the difference in orientation persist? It's unlikely that we will ever know. However, perhaps natural selection favours the capacity for the species as a whole to approach from either side, just in case conditions favour one orientation over the other in the future. Or perhaps it is simply an artefact of the birds' past.

Either way, for now we have the two leanings – and just another reason to marvel at this supremely adapted bird.

Opposite: The lower mandible of this individual twists to the right of the upper. This means that it will work on a cone on its right side, using its right foot to hold it. Fifty per cent of individuals are right footed/billed and the rest orientate to the left.

EURASIAN OYSTERCATCHER

The oystercatcher's trade unions

At first sight, the Eurasian Oystercatcher (*Haematopus ostralegus*) would appear to introduce simplicity into the world of shorebirds, or waders. When you stand on the shoreline, contemplating an estuary dotted with the teeming shapes of tricky-to-identify forms such as sandpipers and plovers, it is a relief to encounter a simple, unmistakable bold, black-and-white form. Oystercatchers are larger than most other shorebirds, too, with pink legs and a splendidly long, thin orange bill which raises faint suspicions that it must be borrowed from another bird from a more colourful environment. Oystercatchers are noisy, too, their loud whistles adding an emphatic sound-track to an unequivocal persona.

The Eurasian Oystercatcher's role on the mudflats would seem to be equally black and white. Early in our birdwatching life we learn that waders with long bills probe into the mud for worms, crustaceans or molluscs, while shorebirds with short bills use their sense of sight to detect much the same food from the surface. In this way the estuary's plentiful resources are partitioned off, with different birds using their different-length bills, with different curvature, for different means. Oystercatchers have long bills, and so their case is open and shut. They will be probing into the estuarine ooze.

It only takes a short time to debunk this theory. With their big bodies and relatively slow rate of feeding, oystercatchers are satisfying to watch. There is a sense of vicarious satisfaction in watching a large, hard-won portion of worm or shellfish disappearing down an oystercatcher's gullet. But you soon see that not every oystercatcher is probing. Indeed, different oystercatchers seem to do different things. Some seem to be surface feeders, not probers, and different individuals don't go about doing things in the same way.

Since oystercatcher-watching is easy for the amateur, you won't be surprised to know that it has been undertaken by scientists, too, and for many years. They have watched the birds with great care over the course of many winters, and colour-marked individuals to find out who does what. And, as always seems to happen with science, they have unmasked a surprisingly complicated situation.

Opposite: This oystercatcher has successfully opened up a mussel by 'stabbing', catching the animal unawares and slitting the adductor muscle that holds the bivalve's shells together.

What they have found is that, faced with the same broad diet, individual oystercatchers specialise. There are three main 'trades', or guilds, each with a different approach to handling estuarine produce. One group specialises in worms, and uses their long bills to probe down into the mud for them, using a combination of touch receptors on the bill to locate the bodies, and sight to read the signs on the mud itself – these, the 'Probers', are the individuals that live up to what you might expect for such a long-billed shorebird. The second 'guild' specialises on shellfish such as cockles and mussels. These hunters tend to forage in soft mud and are on the lookout for molluscs that are filter-feeding with the two valves of their shell open, if only slightly. When they come across them, they lunge forward and attempt to sever the shellfish's adductor muscle, the one that slams the bivalve's shell shut, so that it is useless, and the shellfish helpless. The bill, which is very narrow in lateral section, acts like a letter opener in that it can prise into fractionally open shells. Once the adductor muscle is snipped, the bird can easily cut through the muscles holding the shellfish body to its shell, and shake the morsel free. Oystercatchers using this method are known as 'Stabbers'.

The final group of foragers are, the 'blue-collar workers' among oystercatchers, all picks and shovels – or at least, hammers. They feed over mussel beds, and once they have selected one shellfish that looks tasty (they can tell if a mollusc is diseased), they rip it from its solid mooring and take it to hard substrate, whereupon they unceremoniously hack at the shell until it gives way. The birds aim at the outer rims of the shell, where it is weakest, and different individuals acquire a habit of striking at one particular side or the other. Once they have broken in they, too, aim to sever the adductor muscle, and finish a hatchet job with a smidgeon of finesse. Notwithstanding the expertise involved, this group are known as 'Hammerers'.

To what extent these methods are ingrained within individuals is betrayed by the fine definition of their bill shape. Probers have long, thin bills with sharp tips, Stabbers have chisel-shaped bills, and Hammerers have heavy, blunt-tipped bills. The bill is the result of the feeding method, not the other way round. The end of the bill grows at 0.4mm per day, so can be moulded by wear. Birds that were forced to change their feeding method experimentally were able to do so, and their bill shape gradually changed.

What is it that puts an individual oystercatcher into a particular 'trade union'? The first answer is its parents. In contrast to most other

shorebirds, in which chicks quickly feed themselves without parental help, young oystercatchers are dependent on their parents in their early months, at first for providing all their food, and then in tutoring them on the mudflats. A young oystercatcher may continue in its apprenticeship for up to 26 weeks. Not surprisingly, if its parents are Stabbers, it will learn to stab – nobody has ascertained what happens if the parents are from different guilds – and its initial leanings will be towards stabbing. It will probably become a full-time Stabber.

However, it isn't quite as simple as that. On the whole, females have longer bills than males and this nudges them towards being Probers; the same applies to juveniles, which have more Probers in their ranks than occur in the adult population. There are also intermediates, which presumably means birds that are changing their preference, or interchange methods as the fancy takes them. So the situation isn't black and white.

In fact, there is a fourth method that oystercatchers use to procure food. This method is theft, and it is frequent in oystercatcher flocks. The thief watches other individuals, waits until all the work is done, and then mugs a victim by physical intimidation and loud calls. In a way, stealing is a trade in its own right, and there are oystercatchers that specialise in it, every bit as much as their colleagues practise their own trades.

And who knows, maybe muggers learn their trade from their parents? Questionable practices often run in families.

Above: The Hammerer's bill tends to be blade-like, with a blunt tip.

●

AFRICA

SUNBIRDS

Hovering might be catching

Hummingbirds are box-office, smash hit birds. They feature in innumerable TV documentaries, featuring star-struck scripts. Variously described as jewels and gems, and even possessing names such as brilliants and sunangels, hummingbirds scintillate. Among their glittering cast are the world's smallest birds, those with the fastest wing-beats and the only birds that fly forwards and backwards with the same control. Their flight is so polished that they can attain speeds in excess of 100km/h.

While hummingbirds are A-listers, far fewer people have heard of sunbirds (Nectariniidae). But it's fair to assume that, if hummers didn't exist, sunbirds would assume something of the mantle of the midgets. There are many similarities between the families. Both specialise in sipping nectar from flowerheads. Both families are brilliantly coloured, with many sunbirds outdoing hummingbirds in glittering iridescence and eye-catching patterns. Both live mainly in warm regions, the hummers in the New World and the sunbirds in Africa and tropical Asia. There are numerous species of both (350 hummingbirds and 130 sunbirds), although the sunbirds lack the bill variety that has coevolved between hummingbirds and the plants they pollinate.

There is one missing aspect of their behaviour that keeps sunbirds in the shade: they have not evolved to hover with the effortless brilliance of hummingbirds. Instead, they usually feed when perched, leaning into the flowerhead and sipping from a standing or clinging position. As a result, they have bigger feet than the reduced ones of hummingbirds, and are also larger and heavier overall. Their wings are like those of 'normal' birds, with a longer humerus compared to a hummingbird, but overall their wings are shorter. Their lack of hovering expertise is shared by the third great family of nectar-feeders, the honeyeaters (Meliphagidae) of Australasia.

And this tendency towards perched nectar-drinking is mirrored, to a remarkable extent, by the shapes of the flowers that the birds are evolved to pollinate. In the New World, where hummingbirds reign supreme, many families of flowering plants produce flowers that are oriented outwards towards open space, and lack perches. On the other hand, such plants are fewer, although not entirely absent, in the Old World, with some actually thoughtfully providing perches for the pollinators.

Of course, the situation is never quite as simple as that, because there are plenty of examples of 'lazy' hummingbirds. Although all species can hover,

some prefer to perch and some are specially adapted to use their specialised sharp bills to snip through the long corolla-tubes of highly adapted plants to 'steal' nectar. Experiments have shown that hummingbirds prefer to use perches if these are artificially provided and it is thus the plants that are 'making' them hover. And, as it happens, some birds in the Old World do indeed hover in order to reach nectar. So far, this habit has been recorded about 100 species, the vast majority of which (63 species) are sunbirds. Their hovering lacks the control and expertise of hummingbirds, and they cannot hover for extended periods of time.

In the last few years a fascinating meeting of Old World and New World has played out in South Africa, serving to demonstrate the difference between the continents and pointing to potentially interesting developments in the future. It has been caused by the invasion into Africa of a small South American tree, the Tree Tobacco (*Nicotiana glauca*), that is normally pollinated by hummingbirds. The presence of this newly introduced plant,

Above: Researchers have recently discovered that the Northern Double-collared Sunbird is a regular hoverer at nectar.

which is rich in nectar, has, it seems, set some of the locals a-flutter. Unable to resist the source of food it provides, they have taken up hovering, in the same way a human dad in a midlife crisis suddenly discovers the gym, and cannot keep himself away.

The birds concerned are two species, the Malachite Sunbird (*Nectarinia famosa*) and the Dusky Sunbird (*Cinnyris fuscus*), of which the former seems particularly attached to the blooms of *Nicotiana*. A study by Sjirk Geerts and Anton Pauw of Stellenbosch University found that some Malachite Sunbirds in the Northern Cape region depend on the yellow tubular flowers for much of their winter food, for months on end. And, being unable to reach the nectar easily any other way, for 80 per cent of the time they actually hover

Above: The Malachite Sunbird normally sucks nectar while perching, but in recent years some populations are learning to hover.

in front of the blooms to drink on the wing (Dusky Sunbirds manage 40 per cent). Although they don't hover easily, they can nevertheless hold their position for long enough to make the effort worthwhile. One female Malachite Sunbird hovered for 30 seconds and managed to visit eight different flowers.

The South African study is the first to detect incidences of hovering in sunbirds that amount to something more than the occasional foray. While previously hovering was rare, the skill is clearly now part of a lifestyle. The availability of *Nicotiana* in the Northern Cape, where no native blooms are available in the arid months, has, for example, altered the migration of these birds. At the study site, Malachite Sunbirds used to migrate away in October, but the presence of large patches of *Nicotiana* has enabled them to stay until the end of November. The birds are also most abundant in this part of the Cape where the Tree Tobacco grows, so it has affected the local distribution of the species, too. A bird/ flower relationship has developed. *Nicotiana* trees left open for sunbird pollination produced three times more seed than those experimentally netted to prevent it. Each is benefiting the other. The Malachite Sunbird has taken the place of the hummingbirds that typically pollinated the plant in its native Argentina.

This change in behaviour in Malachite Sunbirds may be local for now, but should *Nicotiana* and other tubular flowers continue to do well in South Africa, there is no reason why the habit should not catch on among other species of sunbirds over a wider region. Perhaps it already has.

It also poses the question: why has the coevolution of richly nectar-packed plants and African birds not led to widespread hover-feeding, as it has in the Americas? Until recently it would appear to be just a quirk of biological history. But now researchers in Cameroon have discovered that a plant called *Impatiens sakeriana* depends entirely upon the Cameroon Sunbird (*Cyanomitra oritis*) and the Northern Double-collared Sunbird (*Cinnyris reichenowi*) to pollinate it, and in the majority of cases this is when they are hovering. If hover-feeding has indeed always happened in this corner of Africa, and the Malachite Sunbirds are showing the ability to follow suit in South Africa, then it is something of a mystery as to why it hasn't caught on more widely.

And for that there is no easy answer.

•

OSTRICH

The benefits of sharing a nest

The Ostrich lays the world's largest egg, a formidable single cell. It measures anywhere between 14cm and 17cm long and 11cm and 14cm wide and it weighs between 1.3–1.9kg, equivalent to nearly 1,000 full adult Bee Hummingbirds (*Mellisuga helenae*). The eggs of Ostriches (*Struthio camelus*), as opposed to being cryptically coloured in common with most ground-nesting birds, are a confident whitish colour, quite easy to pick out in the dry bush country where these flightless birds breed.

You might think that Ostrich eggs, for no other reason than their sheer volume, might be a heavy investment for a female. In fact, the eggs of Ostriches are the smallest, in terms of the ratio of volume to body size, of any bird in the world. They take only about 50 minutes to lay. And Ostriches produce plenty of them; a typical clutch will be 20 eggs, and the record-breaker was 78, and they can have the look of being strewn around. That is an impressive collection of biomass that you would expect the parents to fight tooth and nail to protect. Once again, though, Ostrich behaviour tends to confound expectations. The parents do protect the eggs: they will kick at animals in defence of the nest and they incubate with care, the female taking the day shift from about two hours after dawn, and the male taking the night shift an hour before sunset. But the figures suggest that they don't do it very well. Various studies have uncovered horrendous levels of predation; in some, 90 per cent of the nests were lost during egg laying or incubation.

An Ostrich nest is, not surprisingly, a predator magnet, with all that protein going begging. Egyptian Vultures (*Neophron percnopterus*) have evidently fed on the eggs for so long that they pass on a hereditary capability to open them, using the technique of dropping small pebbles to crack open the shells. Spotted Hyaenas (*Crocuta crocuta*) and Black-backed Jackals (*Canis mesomelas*) are major nest-robbers, at least in east Africa. All of these predators are adaptable and canny.

You might expect that Ostriches have a mechanism to reduce or offset such alarming losses. And you'd be right, but it probably isn't in the way that you'd expect. As it happens, these birds have a very unusual breeding system and a still more peculiar, indeed unique, arrangement of eggs in the nest.

The eccentric breeding system begins when, in season, male Ostriches in an area begin to set up territories. They flag their large, fluffy wings, holding them out to the side and shivering them, they chase other males and,

in a pretty unsubtle invitation to the females, they show off with abandon something that most other birds don't have – a penis. In the breeding season it turns red. Having staked a claim, they also make loud booming sounds and, if all goes well, each male will have his own territory of about 16km², albeit with flaky borders.

A crucial stage in breeding then occurs. Females visit the nest and, if they like what they see, one of them lays an egg. This seals something of a pair bond, and the male and female will now act together to protect the clutch. This, though, isn't the end of the story. Once the female has laid four to five eggs, she then allows other hens in the area to do the same, in her nest. The first female, which is now called the Major Hen, lays on alternate days, while the other females, the Minor Hens, take their turns to lay on gap days. There may be several Minor Hens, perhaps four or five, although 18 have been

Above: At the start of the breeding season, male Ostriches
set up territories and compete to maintain them.

recorded laying in a single nest. These Minor Hens don't seem to be drawn to any particular nest, and most will lay in several.

The end result is that on average a nest will contain between five and 11 eggs of the Major Hen and perhaps 10 from Minor Hens. Although it is a shared nest, only the Major Hen and the male take any part in incubation.

The obvious question then is: why would the pair even tolerate, let alone incubate, the eggs of other individuals in the nest? All the evidence suggests that they are aware of the intrusion, so why don't they act on it?

It turns out that they do act. Researchers have discovered that the Major Hen is able to distinguish between her eggs and those of the Minor Hens. As incubation begins, the Major Hen makes a few adjustments to the organisation of the nest. She makes sure that her own eggs are towards the centre of the clutch, and rolls the eggs of the Minor Hens towards the outside, or even right away from the nest so that they cannot be incubated.

*Opposite: Both sexes incubate the eggs. The Major Hen tends them by day, as here, and the male by night. **Above:** The ring of eggs around this nest are all discarded eggs laid by Minor Hens. Those in the middle are likely to be those of the main female carer, the Major Hen.*

Of course, there is potential benefit here, known as the Dilution Effect. If a predator arrives at the nest and takes a single egg from a clutch of 25, the chances of it being the Major Hen's egg are reduced, the more Minor Hens' eggs there are. And if the predator performs a smash and grab raid, it is far more likely to take an egg from the outside of the pile than the inside – so the Minor Hen's eggs can act as a buffer against predation. Either way, allowing Minor Hens to lay their eggs is an advantage to the territory-holders.

However, there is also an advantage for the Minor Hens. Even if just one of their eggs is incubated to hatching, that is a success; and of course, if a given Minor Hen lays in several nests, it will increase her chances of reproduction in the course of a season.

At first, using strangers' eggs to protect your own would seem a logical and understandable reason for sharing nests, but recent research has shown that it is not as simple as it seems. It was discovered that nearly 70 per cent of eggs in a given nest did not come from a union of the Major Hen and the male; and they also found that individuals acting as Minor Hens at one nest were often Major Hens at another nest, and vice versa. They also measured the kinship between Major and Minor Hens and found that it was not significantly different from the population average. To some extent, therefore, it seems that the population of hens at large share their nests, perhaps with a touch of altruism.

Ostriches suffer from a highly skewed sex ratio, typically at least 1.4:1 and sometimes as high as 3:1 in favour of females. Whether the unusual breeding system is an adaptation to this, allowing a higher percentage of females to lay eggs than would otherwise be the case, isn't known. It seems that there are years of fruitful research into Ostrich behaviour ahead.

Opposite: Once the eggs have hatched, the young are protected mainly by the male. Despite its efforts, only a small number survive to adulthood.

STRAW-TAILED WHYDAH / PURPLE GRENADIER
The strange case of the avian stalker

A quick glance at the photographs on this page and you might think they depicted two colourful and eye-catching African birds, with nothing striking in common. You might guess, correctly, that they were both seed-eaters, just by looking at their conical bills. You wouldn't be at all surprised to learn that they came from the same part of east Africa and lived in similar habitat. But you wouldn't guess that there was much deep connection between them: one is a small finch-like species called the Purple Grenadier (*Uraeginthus ianthinogaster)* and the other is the Straw-tailed Whydah (*Vidua fischeri*). They are not closely related, but instead are in completely different families, the Estrildidae and Viduidae respectively.

However, there is a link between the two unrelated species, a very profound one. Their lives are tangled in a way that is far-reaching and has critical implications for them both. The connection is bizarre, and in its way decidedly sinister, and it shows just how much one species can become intertwined with another.

We should first be clear that, for its part, the Purple Grenadier could easily go about its business without regard to the Straw-tailed Whydah, so far as we know. A whydah-free existence would almost certainly be preferable. On the other hand, the whydah lives almost entirely as the grenadier's shadow. Its life is partly its own, but much of its biology follows that of the smaller bird.

Take its singing, for example. The Straw-tailed Whydah does have its own song, which is an undistinguished mixture of squeaky and harsh notes, but it also sings a perfect copy of the Purple Grenadier's song, and it sings it a lot. When a male is attempting to attract a female to its territory, it spends an inordinate amount of time mimicking the Purple Grenadier rather than singing its own phrase. The female Straw-tailed Whydah clearly responds to this. She is attracted by the mimicry of the male; for her part, she learns the song of grenadiers when still a juvenile.

So far, a little odd. On the other hand, a great many birds all around the world use mimicry to embellish their songs. The secret is to incorporate

*Opposite above: Despite not being closely related to the Purple Grenadier, the Straw-tailed Whydah is entirely dependent on the smaller bird for its survival. **Opposite below**: The Purple Grenadier is a common East African bird, occurring from central Ethiopia south to Tanzania.*

a copied sound into your own species-specific template. It's like putting quotations in an essay: they improve your text, but the overall work is your own. So, maybe the whydah just isn't very good at imitating anything other than Purple Grenadiers? It sings its own song, then mixes in its own favourite grenadier riff – just rather often.

But such an innocent suggestion is undermined by the detail. It happens that the Purple Grenadier's rather pleasing, rising and falling song varies from male to male and from place to place. And guess what? So does the whydah's, in precise geographic synchrony with the grenadier's. That suggests that something intentional, rather than incidental, is going on. It seems that the Straw-tailed Whydah is compelled to imitate the grenadier as a matter of necessity.

And it isn't only the territorial song that the whydah imitates. It also imitates the Purple Grenadier's alarm call, excitement call and several of its contact calls, and incorporates these into its singing bouts. No other species is treated as a vocal template. And when do you think the Straw-tailed Whydah starts singing its song? It isn't at any random time of year. It happens, of course, when the Purple Grenadiers themselves start breeding. This is the first clue that explains their peculiar and overbearing relationship to the Purple Grenadiers. The male whydahs moult into their breeding plumage and become fertile at the same time that their hosts do, in readiness to overlap their breeding season in the very fullest sense. By now you might have cottoned on to the fact that Straw-tailed Whydahs are brood parasites of Purple Grenadiers.

Thus, a female Straw-tailed Whydah lays eggs in the nest of the smaller bird. Her eggs are of the same basic colour, although they are slighter larger and rounder than the grenadier's. Nonetheless, they are accepted, quite possibly even when a female grenadier is aware of the intrusion. Unlike the

Common Cuckoo, the most famous brood parasite, the whydah does not destroy the host eggs, but simply pops her own egg in amongst those of the host. Sometimes it lays a single egg, sometimes more, but there is no damage to the nest or clutch of the host, at least at this stage.

Little is known about any struggle between the two sets of nestlings. The young whydah do not drive the rival nestlings out, but studies on closely related species suggests that the parasites do hatch a day or two earlier than the rightful nestlings. If this is the case, you might expect the parasites to out-compete their rival chicks for food offerings, although this hasn't been proven as yet.

What is in no doubt is that, while Straw-tailed Whydahs are not particularly good egg mimics, their chicks are absolute masters of the art: here again they shadow the host almost perfectly. The markings on the mouths of the parasites – in this case orange and blue, with a white-edged tongue – perfectly match those of the hosts. And the nestlings themselves are little actors, going as far as to mimic the swaying movements of Purple Grenadier youngsters. The deception continues into the first few weeks of juvenile life, when the parasites remain with their foster families. Only after a couple of weeks do they join flocks of their own kind.

This particular example of avian stalking is not unique to the Straw-tailed Whydah, but also occurs in other closely related species, which have different hosts, albeit all of them seed-eaters. The big question has to be: why have they evolved so specifically to depend on this single host? Why don't they use a variety of different hosts, like a cuckoo? Why have they evolved such a complete and complex fabric of behaviour just to ensure that their lives entwine with the host?

And, more pertinently, why don't they just build a nest for themselves?

BOUBOUS

It takes two

Often you don't have to speak wisely to be heard, you just have to speak over and over again. Repetition, rather than quality, gets the message through. The same applies to a human appreciation of bird sounds. When you visit a new place there may be any number of stunningly euphonious songs making up the soundscape, but the ones you will remember will be dominant and repetitive. In Africa, for example, there is the endless droning of doves, always somehow making it feel hot; and in most parts of Africa there are also boubous, bush shrikes that are members of the genus *Laniarius*. With their loud, tuneful whistles, variously described as gong-like, liquid or sweet, they lace the air with a tropical feel.

The songs of boubous are remarkably persistent, and they are among the few bird sounds that ring out through the heat of the African day. The Slate-coloured Boubou (*Laniarius funebris*) of central east Africa may sing 260 songs an hour during the afternoon, and several species are known to deliver more than a thousand phrases between sunrise and sunset. If you wanted an example of attention by repetition, this would be it. The songs are far from complex: in the case of the Yellow-crowned Gonolek (*Laniarius barbarus*) of West Africa, there is typically a liquid note that rises and falls, and could be rendered 'Oriole', together with a rasping 'trrk'. And as they say: 'That's all, folks'. Each song lasts less than a second.

Boubous aren't unique in repeating their songs interminably. North American birders will be familiar with the famous occasion when a Red-eyed Vireo (*Vireo olivaceus*) was counted singing its short ditty 27,000 times in a day, and in Europe male Yellowhammers (*Emberiza citrinella*) have been known to repeat their song 5,000 times. But there is a major difference. In the case of vireos and Yellowhammers there is just one bird singing. In the case of boubous, the song is a duet between two members of a pair.

When you hear the song in the wild, it is hard to believe that two birds are making it, because the voices are so precisely co-ordinated. In the case of the Tropical Boubou (*Laniarius major*) and Slate-coloured Boubou, one bird follows the other without any discernible hesitation, while in the case of the Yellow-crowned Gonolek the two parts actually overlap. The response time between one bird's song and another is less than a tenth of a second. In the field, unless you are listening very carefully and the birds are spatially far apart, you simply cannot tell that it is a duet at all.

This degree of expertise might be expected if one bird always started the same way and the other always did the agreed response. But this isn't the case. True, it is usually males that utter the liquid notes and females that make grating or harsh notes in response, but sometimes this is reversed, with the female making the male-like whistle and the male responding. Furthermore, each pair has a variety of different motifs that it will utter in different situations or for different reasons. Despite this, there is no loss of synchrony; it always sounds just like the one bird.

The degree of co-ordination required to perfect the duet is impressive, and the sheer number of songs given by the pair is startling. Nobody could doubt that the songs of boubous serve an important function. But so far, there is one question that scientists are struggling to answer. Why do boubous and

Above: The endlessly repeated song of the Yellow-crowned Gonolek lasts less than a second, but it is performed by the two members of a pair.

other duetting birds use two voices, when in most parts of the world the song is only sung by a male? Red-eyed Vireos and Yellowhammer males manage to keep their territorial borders intact without resorting to help from a female, so why can't boubous?

There is no clear cut, catch-all answer to this question as yet, although there are plenty of hypotheses. And of course, there may be several functions behind duetting, rather than just one. The fact is that, while many birds of different taxonomic groups all around the world take part in duets (about 400 species in all), the vast majority don't. Duetting may be essential for boubous, but it isn't essential for most birds.

There is no doubt, however, that pairs of boubous do duet for territorial reasons, just as single birds do. Experiments in which a neighbour's song is played back to a pair on territory invariably cause a steep and rapid increase

Above: When members of a pair of duettists such as Slate-coloured Boubous sing, the degree of synchrony in the duet may reflect the strength of their relationship.

in the rate that the incumbents repeat their songs. Recent research on tropical boubous has revealed that they have a special song that is only given when they have driven away rivals. It is never heard in any other context, so you could describe it as a triumphant duet (or a taunting duet towards the neighbours). In the context its use in a duet is appropriate, since both partners collaborate in excluding rivals from their territory.

However, the sheer regularity of duetting, with hundreds of songs a day uttered throughout the year, suggests that it cannot only be for territorial purposes. It seems highly likely that the birds are passing messages to one another. Duetting could be used to ensure that a partner has a bird's full attention and is not off on a dalliance with an intruder of the opposite sex. Every time it responds to its mate, its whereabouts are automatically checked. This is a form of mate-guarding.

The degree of synchrony within a duet could easily form a variable within the pair bond, too, which could be an indication of one partner's commitment to the other. Fast, regular responses, for example, could signal a high degree of commitment, while hesitant responses, or no response at all, could signal a potential split in the relationship. When only a male bird sings, as in a Red-eyed Vireo, a listening female will have no way of measuring the male's commitment to herself, but only to his territorial defence. Duetting could be a very reliable measurement of this commitment. A wavering in duet co-ordination would not only be evident to the members of a pair themselves, but also to their neighbours, potentially compromising their relationship, and their grip on the territory. Indeed, a study of Magpie-larks (*Grallina cyanoleuca*) in Australia found that the degree of precision in duets was a measure of how able and willing a pair was to co-operate for the defence of their territory.

And speaking of co-ordination, it has also been suggested that duetting enables the two members of a pair to synchronise their breeding roles, so that the correct hormones are released in synchrony, making a successful breeding attempt more likely. However, since female birds of other species are known to be so stimulated just by listening to male song, it is unlikely that duetting would play an important role. One thing that several of these hypotheses do seem to accept, though, is that despite the high rate of repetition in boubou songs, quantity isn't everything. As far as synchrony is concerned, quality matters as well.

WIDOWBIRDS

A tale of two tails

Researchers into bird behaviour follow fads and fashions, just like everybody else. At the moment one of academia's pet subjects concerns mate choice, asking the question: what characteristics are used by male or female birds to make wise decisions about whom to pair up with? Over recent years many extraordinary criteria have been uncovered that play Cupid, including ultraviolet reflectance on the feathers, the number of parasites in the nest and the number of speckles on a thigh. Some of the modes of selection are very subtle, and some are not subtle at all. Firmly in the latter category are the mating choices of the Long-tailed Widowbird (*Euplectes progne*) of southern Africa.

In the case of this species, as it is in most species of birds which have been studied up to now, the 'choosing sex' is the female. It is therefore incumbent upon males to dress up glamorously, so to speak, to display well and, where necessary, to sing with vigour. In this species the stakes are particularly competitive, because males are frequently polygynous, meaning that a given male could in theory mate with any number of females. The other side of the coin is that if a male is not up to scratch – or at least, if the local opposition is of higher quality than he is – he could very easily go an entire breeding season without mating with anybody. It could be a winner-takes-all scenario, with the most successful males acquiring a harem of up to five females.

What, might you ask, do female Long-tailed Widowbirds look for in a male? There is one glaringly obvious candidate: the tail. In the breeding season the smartly attired males do indeed sport remarkably long tails. They are also broad, with the central tail feathers the longest and the outer feathers progressively shorter. Black and silky, these tails are so elongated that they impede flight. The body of the male is no more than 15–20cm long, while the tail can be as long as 50cm. During the breeding season the males have an eye-catching display in which they fly with slow, laboured wing-beats, sometimes hovering, and dangle their tail down, sometimes also raising the body so that the tail forms a sort of 'keel' below it. The display shows off the

Opposite: No matter how much a male Long-tailed Widowbird ruffles its feathers and shows its epaulettes, it seems that the length of its tail matters the most.

Above: The tail impedes a male's flight —and that is partly the point.

tail to advantage, but while the tail is the obvious sexual character, is it this that the female falls for?

The first piece of supporting evidence in favour is that different individual male Long-tailed Widowbirds have tails of variable length, so tail length is a candidate for female choice in the wild. They also have red epaulettes, which they flash coaxingly, and they sing with gusto, so there are other possibilities. Proving the link required a series of neat experiments in tail manipulation, carried out by Malte Andersson and his co-workers back in the early 1980s.

The scientists went into the widowbirds' short-grass habitat and caught a large number of males. They colour-marked them and divided them randomly into three different categories. In the first, control group they cut the tails about half way down, and then stuck them back in exactly the same place, so that they were not altered in length. The second group, however, had to cope with some drastic tail surgery. About 25cm of the central section of their tails was removed, and then the tip was reconnected to the base, meaning that the tail was shortened by almost half its length. The third group also had their tails cut, but this time the central segment of a member of the second group's tail was glued in between the base and the tip, thus radically increasing the length of the tail. The males were then released back into their territories, to display and attract mates. How would the females react?

The results were interpreted by the number of new nests found in a male's territory after surgery; since the females build the nests, each is a measure of a newly attracted individual. (The females, incidentally, incubate the eggs and rear the young alone, making the selection of a high-quality male the only thing to worry about – not his parenting skills). On average, on the control territories one new nest was found; where birds had their tails shortened, on average less than half a new nest appeared in the territory. On the territories where the males' tails had been lengthened it appeared that all the male's Christmases had come at once, because on average they were able to attract two new females each. In short, the results are pretty spectacular; females were clearly swayed by the experimental modification of the tail, strongly suggesting that they prefer males with longer tails. In the wild, the birds with the longest tails are likely to be older, more experienced birds, which are always the most attractive in bird populations.

At first sight, the pure empirical length of a male's tail would seem to be a somewhat shallow way of measuring his suitability, and perhaps one that isn't reliable. However, there are a number of ways in which a long tail might be an expression of a bird's inner vigour and fitness. For one thing, the tail is a clear impediment to flying, making day-to-day getting about hard work, let along escaping from predators and performing displays. A longer tail is likely to be a greater impediment than a shorter tail. So, by growing a long tail, a male is giving out the message that it can cope with everything despite its tail. Males with shorter tails have less in the tank than longer tailed birds because if they had the same amount, they would use it to grow longer tails. It is also possible that the act of growing a long tail is particularly energy-demanding. If that was the case, it is simply cause and effect, so clearly the better males would grow longer tails.

In contrast, there is another species of widowbird that may sometimes be found in the same places as the Long-tailed. The lifestyle of the Fan-tailed Widowbird (*Euplectes axillaris*) is very similar, with the polygynous males attempting to attract females into their territory – some have been known to woo eight different females – with flight displays, showy epaulettes and song. Furthermore, in this species, it has been shown with similar experiments that females once again prefer males with longer tails. The difference in this case is that the tails of male Fan-tailed Widowbirds are barely longer than the bird's body, so much so that they probably don't impede the bearer's everyday life very much. In this case, perhaps the tail tells a different story, but what it might be has yet to be determined.

***Opposite:** The Fan-tailed Widowbird has a very similar lifestyle to Long-tailed Widowbird, and though its tail is much shorter, the length itself is still important.*

ASIA

GREATER RACKET-TAILED DRONGO
The life of a professional agitator

The Greater Racket-tailed Drongo (*Dicrurus paradisaeus*) is the sort of bird that demands attention – not just in life, but during perusals of lists of species, as well. It's a medium-sized bird, smaller than a crow, but with a long tail with outermost feathers that are greatly extended and denuded of barbs along much of their length, except for spatula-shaped tips. The tail ornaments trail behind the bird in flight in an out-of-control fashion, with minds of their own. The plumage is glossy-black, including the crown feathers which are somewhat unkempt, like recently washed hair. The beady eye is red, and the bill short, strong and curved downwards. By this description, if you think you have a character, you are right indeed. And if you think that the Greater Racket-tailed Drongo is not to be messed with, you are right again.

This bird occurs in forested habitats over much of Asia, and is quite easy to find because it is usually associated with flocks of other birds. Indeed, it seems as though flocks are its lifeblood. In the morning a Greater Racket-tailed Drongo will often act as a 'nucleus', summoning other potential flock members by making loud calls, including mimicking the sounds of the species it wishes to attract. Thus, in some parts of its range it particularly seeks out the Ashy-headed Laughingthrush (*Garrulax cinereifrons*) and the Jungle Babbler (*Turdoides striata*), both of which travel around in parties of their own. So attracted is the drongo to these species that it makes a bee-line for them, imitating their songs and calls and adapting its own perching height to be closer to individuals of their species, rather like an unhealthily obsessed fan. Something similar happens in Sri Lanka with another flock forager, the Orange-billed Babbler (*Turdoides rufescens*), and in Malaysia it seems that the irresistible attraction is woodpeckers. Wherever it goes, it seems, the Greater Racket-tailed Drongo prefers to have companions.

There is a perfect logic to its strategy, for the drongo is a commensal forager – in literal terms, this means that it 'feeds off the crumbs off another's table'. In practice the bird fields insects which have been flushed by its 'beater' colleagues busily foraging in the foliage below. Being a perch-hunter rather than skulking in vegetation, the drongo strongly benefits from

---

*Opposite: The Greater Racket-tailed Drongo is an
irrepressible, noisy and common bird of Asian
forests. It perches still to feed, waiting to pounce on
something flushed by feeding birds below.*

keeping a close watch on the behaviour of smaller flock members, because they will consistently uproot insects from their hiding places. Sometimes, indeed, the drongo will follow mammals, too, such as primates and squirrels, including those feeding on the ground.

The drongo's enthusiasm for flocking is generally reciprocated; the invitees flock by choice and the drongo has no power to coerce them. The other birds know that there are advantages for them, too, in having drongos around. The sharp-eyed perch-hunter is quite brilliant, it seems, at spotting predators, so much so that everybody responds to its alarm calls immediately. Even more remarkably, the Greater Racket-tailed Drongo will sometimes give species-specific alarm calls, as if it were a special guardian to its favoured individuals. And that's not all, for drongos may act not just as sentinels, but actually attack predators as well. As a group the drongos, of which there are about 20 species, are remarkably bold and aggressive, and seem unfazed by the size of the animal that they are mobbing. Large birds of prey, crows, monkeys and even people are sometimes physically attacked. Remarkably, a Greater Racket-tailed Drongo has been recorded actually 'riding' on the back of an enormous Great Hornbill (*Buceros bicornis*) while pecking at it during a sustained assault (the hornbill is not a bird of prey as such, but could possibly take the occasional egg or chick). With such a high level of protection, it is hardly surprising that the forest birds are content to have drongos in their midst.

However, the heroic nature of these birds should not be over-estimated. Even the most admirable characters have flaws, and at times give in to temptation. And the very same is true of the Greater Racket-tailed Drongo; sometimes, it should be admitted, it does abuse its relationship with other flock members, and it does so with a very neat trick of deception.

Imagine that, as a drongo, you are having a bad day. Perhaps you keep missing insects that have been flushed towards you or, more likely, feeding is difficult for everybody because of environmental conditions. When this happens, it isn't hard to imagine a bird switching from fielding what another bird has flushed, to stealing from it directly – lots of bird species do this routinely. And since drongos are notoriously aggressive, and often larger than their flock colleagues, it is remarkable indeed that in Sri Lanka, for example, piracy accounts for only three per cent of their feeding attempts. More surprisingly, the way they steal is neither aggressive nor physical.

●

What they do is provide a distraction. Sometimes an aggressive call stops a bird in its tracks, allowing the drongo to pick up the food that the startled bird has dropped. On other occasions the vocalisation is apparently deliberately deceptive: the drongo spots a particular species carrying a large morsel of food and then mimics the alarm call of that very same species, all the while with no predator in sight. This once again causes the smaller bird to flee, leaving the food behind. It is a ruse to snap up a stolen meal.

Quite obviously, however, a false alarm sets a precedent which, in the end, could be to the drongo's detriment. Too much repetition of the ruse is likely to backfire, in that the victim latches on to the deception, fails to fall for the mimicked calls or, worse, eventually becomes reluctant to be part of the drongo's precious flock at all. Deception can only be used sparingly, however effective it might be.

The truth is that the Greater Racket-tailed Drongo is, for the most part, a benefit to a small foraging bird. The odd piece of theft and deception is a small price to pay for its protection racket.

*Above: Drongos often mob far larger birds, such as this Plain-pouched Hornbill (*Aceros subruficollis*), seen approaching its nest cavity, and will sometimes attack them physically.*

●

YELLOW-BROWED WARBLER

The wrong-way migrant

Bird migration is a wondrous phenomenon and we rightly marvel at it. The seasonal movements of birds have been exhaustively studied for many years, and yet the subject keeps throwing up extraordinary facts and figures. We hear about birds using a magnetic sense to find their direction, of using the sunset and polarised light. We hear about how young birds a few days old watch the movement of the stars from their nests, and learn the rotation. We hear how birds migrate from Alaska to Hawaii, somehow managing to pinpoint this small island chain in the vast ocean, when a small mistake in orientation could send them into the oblivion of treacherous salt water. We hear that even small birds can fly 5,000km without stopping, and larger birds 10,000km. Yet even after these enormous journeys, the same birds can return to exactly the same breeding site year after year, even within a few metres. Still others return on the same day each year, requiring a formidably accurate body clock. We have learned that much of the trip is programmed into birds genetically, so that they can make the journey unaided. The more we learn, the more brilliant migrant birds become.

What people often don't hear about, however, is when birds get it wrong and make life very difficult for themselves. Birders are aware of such things, and they revel in encounters with waifs and strays that have got lost. All kinds of hazards face migrants, including capricious winds, fog, rain and mountains. They conspire to usher birds unintentionally in the wrong direction, too far removed and perhaps too exhausted to recalibrate their famous migratory senses. Such events are minute tragedies for individuals, tiny sample sizes of mishap, easily consigned to evolutionary irrelevance. The same applies to any number of mutant birds, whose inner programmes malfunction and lead them to disaster without the effect of malign weather.

In recent years, a number of cases have come to light of a phenomenon, far removed from personal catastrophe or defect, in which many individuals seem to misfire in the same way. There seems to be a kind of institutionalised incompetence which, by virtue of its frequency, has become a major part of a species' ecology. Probably the best known of these is the curious case of the reversing Yellow-browed Warblers (*Phylloscopus inornatus*).

The Yellow-browed Warbler is a diminutive inhabitant of montane broad-leaved forests on the southern fringe of Siberia, breeding from the Urals eastward to the Sea of Okhotsk. In common with a host of other species, it migrates on a south-easterly bearing in the autumn to winter in warm

regions of north-east India east to Taiwan, and it is a familiar bird in such places as Hong Kong and Singapore. This migratory route should keep it clear of the birdwatching hubs of western Europe to the tune of thousands of kilometres. And yet, bizarrely, there are few birders in Holland and Britain, for example, who have not seen one. The Yellow-browed Warbler is still a rare bird in these parts, but it is by far the most common Siberian vagrant here. On average there are 320 records a year in Britain, and this must be a small fraction of the true numbers involved. If there are so many just in this one country, it is fair to assume that it is a major movement. By comparison, a related species, the Western Bonelli's Warbler (*Phylloscopus bonelli*), breeds in France, a continental neighbour of Britain. Yet this species is much rarer, with only two records a year on average. By rights the Yellow-browed Warbler should be by far the scarcer bird, but it isn't.

Above: The Yellow-browed Warbler is arguably more famous for getting its migration direction wrong – turning up in Europe – rather than following its correct route between the Siberian fringe and south-east Asia.

What, then, is bringing these small birds to the 'wrong' place in such numbers? At the moment we don't know the complete answer, but the question is important, because there are several other birds that turn up regularly in unlikely places, too. However, an analysis of what is going on is decidedly intriguing.

The Yellow-browed Warbler's eccentric vagrancy becomes a little easier to understand when you realise that, if you were to reverse its normal route by 180°, that would indeed bring it to western Europe. The reality is only theoretical, because it would happen only if birds were to follow along the exact shortest route over the surface of the earth, the great circle, disregarding the vagaries of topography and weather, and so far there is no proof that any birds follow the great circle. It is telling, though, that the angle by which they get their migration 'wrong' is close to 180° in real life, taking the birds north-west instead of south-east.

How, though, could such a flip happen? The most persuasive suggestion made so far is that the birds' body clock – or at least, their season clock – is faulty, and they are six months out. Perhaps they think it is spring, when their natural inherited tendency would take them north-west, when it is actually autumn? And perhaps they think it is autumn when it is really spring, in which case they might just arrive back where they started?

Another possibility is that the north-westerly migration is not a mistake, and that a sub-population of Yellow-browed Warblers have been wintering in Western Europe for millennia, without being discovered. As mentioned above, these birds do breed in the Ural Mountains; perhaps there is a migratory divide between western and eastern populations, and western birds have always moved south-west, not south-east in the autumn? Another possibility is that the birds of the Ural Mountains or elsewhere have only recently begun to winter in Europe, owing to a favourable genetic mutation. The occasional Yellow-browed Warbler has been recorded in west Africa in recent years, adding weight to this suggestion.

For now, we don't know what is going on, and it will take many more ringing recoveries and other research to find out. In the meantime, there is something quite charming about a large group of migrants making mistakes *en masse*. It doesn't tarnish the image of the migratory bird. If anything, it makes it still more amazing that so many migrants successfully reach their destinations.

Opposite: It isn't known whether these 10cm long migrants make it back to the breeding grounds after getting their migration 'wrong'.

PHEASANT-TAILED JACANA
Children of the lily-pads

Young jacanas are brought up in an unsteady environment. The eight species of these unusual waders in the family Jacanidae are unique for their permanent attachment to a micro-habitat of floating vegetation, especially water lilies. You do not find jacanas in every freshwater marshland, but only where the surface is partly or completely covered by leaves and flowers. The birds feed, preen and socialise on this thin, if luxuriant skin that wavers over water that can be several metres deep. The birds' primary adaptation for thin-layer trotting is to have unusually long toes splaying out about 15cm across in the case of the Pheasant-tailed Jacana (*Hydrophasianus chirurgus*) – which spread the bird's weight over a wide surface area, preventing the vegetation giving way beneath it. When things go wrong, jacanas can swim or fly, but it is extraordinary how rarely the lily-pads fail them.

Not only do jacanas forage and bicker on this unsteady floating carpet, they also build nests and raise their chicks on it. And while lily-trotting is undoubtedly a great career for a bird that feeds on the myriad of small invertebrates associated with freshwater, it presents its share of problems for laying eggs and bringing up chicks. For a start, jacanas have to build a nest that won't sink. And secondly, they need to be experts at hiding both nests and eggs, which are vulnerable to predation in a habitat on a thin layer without many hiding places. None of this is easy, and as a result, most jacanas, including the Pheasant-tailed, also have an unusual breeding system that makes their parenting skills capricious, to say the least. Or, might we say, unsteady.

The progeny start life afloat. The female builds a nest that is small (14cm diameter), barely protected and wet. It tends to be little more than a pile of leaf-stalks and other vegetation, made on a floating mat, and it has a depression in the middle about 3cm deep, where the eggs are laid. The nest isn't waterproof, so the eggs are at best laid and incubated in the damp, and at worst actually touch the water. Some females simply lay an egg on a water-lily leaf. The eggs are plain and well camouflaged, but have no physical protection at all. The impression is that eggs are relatively cheap currency in a jacana's breeding system.

In fact, they are cheap to the bird that lays them, because female Pheasant-tailed Jacanas practise what is known as sequential polyandry and can be remarkably productive. They pair up with a male and lay their standard clutch of four eggs, and then a few days later they pair up with

another male and lay another clutch, and so on. Some female jacanas have been known to lay 10 clutches in a season. The females are very much larger than the males, sometimes twice as heavy (this form of inequality, known as 'reverse sexual-size dimorphism', reaches its avian apogee in jacanas), and they are also in short supply. The males greatly outnumber the females.

Quite naturally, a female laying multiple clutches cannot incubate them all, and as it turns out, she doesn't incubate any of them, but instead more or less abandons them. Her maternal instincts are principally geared towards egg production, although a female Pheasant-tailed Jacana will sometimes help to protect her young against predators; this, however, doesn't seem to happen very often. Rather, the care of the eggs now passes to the males.

Above: Jacanas are the only birds in the world that live most of their lives on floating vegetation. Their toes are consequently long to spread their weight out.

If you read the literature about the paternal care of Pheasant-tailed Jacanas, you are left with the distinct impression that the observers cannot quite fathom what the motivations of the fathers are, and indeed the males' efforts seem to veer between intense protection and outright neglect. On the one hand they are devoted and assiduous, while on the other hand they can be abusive and worse. Once again, they are capricious.

One of the ways in which male jacanas tick good parenting boxes is in the regularity with which they move their breeding units around, for their own safety. If they deem their nest to be vulnerable to predation, for example after some kind of disturbance, Pheasant-tailed males will simply build another nest in a different location in their territory, and then move the eggs. Having built the new platform, they kneel down at the old nest and collect each egg, one by one, by wedging the pointed end of the pear-shaped structure into their breast, and holding it in place with their bill. They then walk awkwardly to the new nest, travelling up to 11m while performing this balancing act.

Above: A male Pheasant-tailed Jacana incubating. If the male deems the nest site under threat, he will quickly build a new nest and move the eggs, one by one, by carrying them, wedging them between breast and bill.

These translocations happen with surprising frequency: a male in Thailand was observed to move its clutch four times during the 26 days of incubation. This was no mean feat, because it took a full three hours each time.

And yet, despite the preciousness of their clutch, Pheasant-tailed Jacanas are not attentive all of the time. They invariably incubate through the night, and also during the hottest part of the day, 11.00 to 15.00, but for the first 10 days or so they are absent for the rest of the time. This period coincides with the heaviest losses, and why they should do this is a mystery; over half of all jacana clutches are lost before they hatch. Things do improve for the last week of incubation, but by then it is too late for many eggs.

Once the eggs hatch, the males are unequivocal about protecting the chicks, and one aspect of jacana parental care is particularly celebrated. The birds have enlarged bones in the inner wing, and this enables them literally to scoop the young under their wings and run away to safety with them. At other times they simply lead the young away quietly, like a chicken might do. If, however, a predator is nearby, jacanas might directly attack it, or they might attempt to distract it by lying down with their wings spread, as if incapacitated. Apparently some jacanas have been seen to extend one foot and move in a circle, as if the other foot had snagged on a piece of vegetation.

This courageous protection is an example of an extreme commitment to parenting, so it is hard to imagine that this same species of bird would ever do the complete opposite – infanticide. However, male Pheasant-tailed Jacanas have been observed doing exactly that – and with some regularity. They don't kill chicks, but it seems they do sometimes sacrifice eggs.

In a study in Taiwan, a high proportion of male Pheasant-tailed Jacanas were observed to throw eggs out of their own nests, to certain destruction. This might seem shocking, as infanticide always does, but it would seem that these were victims of the jacanas' unusual mating system. Where a female mates with a number of males in sequence, there is a short window of uncertainty about the paternity of certain eggs when a female changes over from one suitor to another. A male's sperm is viable for up to five days inside the female's reproductive tract, and if a new partner has any doubt about the provenance of the first few eggs in the clutch the shared female lays for it, it seems that he takes the extreme option and kills the doubt.

After all, he doesn't want to waste his parenting on somebody else's egg.

ARABIAN BABBLER

Keeping its friends close...

Alarm breaks out in the Middle East. In a patch of dry scrub on the edge of a stony desert, the peace of early morning is interrupted by a series of loud, urgent 'tzwick' calls, which echo across the deeply fissured, red-brown rocks of a dried-up river bed. Knowing that these calls indicate extreme danger, several small, greyish-brown birds with long tails and fine dark streaks on their foreparts hunker down in a dense growth of thorny vegetation, and a Lanner Falcon (*Falco biarmicus*), knowing that it has been spotted, glides harmlessly by. A series of further 'twick' calls and trilling sounds from the smaller birds usher the foiled predator on its way, the avian equivalent of cat-calls and rude hand signals. An everyday victory has been scored by this close-knit group of Arabian Babblers (*Turdoides squamiceps*).

They owe their safety to the vigilance of the bird that first made the alarm call. Sentinel behaviour, in which a bird perches above its colleagues and devotes time to watching their backs, is particularly well developed in Arabian Babblers and their kin, and they have evolved a sophisticated series of calls that helps them react appropriately to danger. Different intrusions induce different types of calls and call repetition, one for a perched predator, one for a distant predator and still another for a ground predator, such as a snake. Once alerted, the flock reacts according to the threat. That reaction may be to keep low and wait for it to pass, or to mob and harass a predator.

But for the Arabian Babbler's system to work effectively, the sentinel must be on its mettle. The survival of each member of the flock depends on the assigned bird's ability to keep a sharp lookout and to set off the vocal early-warning system promptly. Not surprisingly, vigilance isn't down to a single individual throughout the day, but relies on a shift system, with several birds covering the rota. In between, individuals need to feed and rest and presumably, as in the case of humans, they have a finite concentration span.

Sentinel duties are costly. Birds engaged in this activity would be expected to forage less than their colleagues on any given day, and they also place themselves in exposed positions at greater risk to predators. If you were to analyse the place of sentinels in Arabian Babbler society, your first thought might well veer towards an altruistic bent – the sentinels are putting themselves out for the sake of their conspecifics. Indeed, they are risking their lives with little obvious reward.

But Arabian Babbler society is complex and it turns out that being a

sentinel has its rewards. In human parlance being a lookout would be
described as a 'responsible job'. And just as responsible jobs carry kudos in
human society, so does the position of flock sentinel in Arabian Babblers. As
a result, certain members of the flock actively seek the role – and compete
among themselves for it. The bird that begins the lookout stint in the
morning is the group's dominant bird, and this same bird might well give up
three hours from its day for sentinel services. In fact, it will actively prevent
a subordinate from taking the next shift, presumably if it feels threatened.

But why is kudos so important for Arabian Babblers? They live in groups
of 6–13 individuals on average, and within each group is a strict hierarchy.
The oldest male and female in the group are the top-ranked, dominant birds

---

Above: An Arabian Babbler stands sentinel while its fellow flock-members
feed below. Being a sentinel is a highly sought-after activity.

of their sex, and they are responsible for each year's breeding attempt, the oldest female laying all or most of the eggs, and the dominant male siring all or most of the young. Despite being shut out of the breeding attempt in the reproductive sense, all members of the flock contribute to the various nesting activities, including incubation, feeding the young and guarding the fledglings.

Above: Arabian Babblers are highly intolerant of potential predators in their group territory. They sometimes attack and kill snakes.

Within this arrangement, however, there are sub-plots. Whilst a bird cannot alter its rank in the flock – unless it is the oldest it cannot be dominant – it can behave in ways to enhance its social status within the group. For example, if there are three or four males in a flock, then the three subordinate birds can fight it out amongst themselves, day by day, to assume top position. For much of the time, their posturing doesn't amount to much. But during the breeding season, a higher social status can pay dividends. In some circumstances, birds other than the top-ranked individuals do contribute to the breeding attempt. A second-ranked (beta) male might successfully mate with the dominant female and sire offspring, while a subordinate female might successfully contribute to the group's clutch.

The posturing in the ranks below the dominant male also affects this bird's own behaviour, because the individual at the top of the second tier is its immediate threat. Therefore, the alpha male tends to make life most difficult for the beta male, while allowing the less threatening gamma male (and any others further down the ladder) more freedom of expression. For example the dominant male will actively prevent the beta-male from taking sentinel duties, while being content to defer to birds of lower social status.

Not surprisingly, the social status of Arabian Babblers within a group does not just depend upon the allocation of lookout duty. There are many other behaviours that are indicative of, or may alter, the status quo. Just as sentinel duty may be dressed up as altruism, the same goes for preening another individual (allopreening) or feeding a conspecific, bill to bill (allofeeding). In each case, the act of 'serving' another bird by feeding or preening it is in fact an expression of dominance. Every time a bird accepts food or pampering from another, its social status drops.

Why, though, should Arabian Babblers express their status in such a counter-intuitive manner? In a simpler world, couldn't they simply fight it out between them to see who should hold the highest status? It seems that in Arabian Babbler society, avoiding conflict is very important – it does occur, for instance when two groups are fighting over territory, and seems quite often to produce casualties. It seems that, within the group, it is far better to resolve conflicts in a subtle but non-aggressive way.

Thus it is that, as far as sentinel behaviour is concerned, he who watches most from the highest perch, holds the highest perch.

SWIFTS AND SWIFTLETS
Living in the dark

Swiftlets aren't colourful, they aren't large or particularly noisy, or spectacular in any way. The species, of which there are about 23 – give or take some taxonomic reshuffles – are hard to tell apart. They are an Indo-Pacific group, commonest in southern Asia but reaching across the Pacific east to the far-flung Marquesas Islands. And when you do see them, they have a tendency just to fly routinely over forest above the treetops, or even over towns or offshore islands, doing what swifts the world over do, chasing flying insects, which they catch in their gapes, using vision as the main sense.

If you were to keep watching swiftlets even for just for a minute or so, you might appreciate their supreme powers of flight and manoeuvrability. In common with their relatives the hummingbirds (Trochilidae), all swifts and swiftlets (Apodidae) fly with their fingers: the 'arm' bones are greatly reduced, allowing the 'hand' to perform the fine control of their long, pointed wings. Swiftlets accelerate and decelerate deftly, so much so that you can't appreciate how they do it, and their ability to twist and turn in tight spaces sets them apart. If, somehow, you could fit a small video camera upon a swiftlet's back, it would broadcast quite a ride.

If you followed the swiftlet for considerably longer, beyond its day-to-day foraging and into the brief tropical evening, then it would bring you into a far more unexpected realm. Swiftlets breed and roost in caves, tunnels, or even in quiet corners of the interior of buildings. As soon as the day comes to an end, they retreat into a world of permanent, and often total darkness. These diurnally active feeders, whose eyesight is so acute that they can target an individual gnat in the air, enter a realm where they can see nothing at all. Yet they cope by using echolocation, the only birds in the world apart from the Oilbird (*Steatornis caripensis*) of the Neotropics that can do so. Echolocation is the ability to orientate entirely by picking up echoes of self-made sounds.

Swiftlets are unique among animals in combining the ability to echolocate with an essentially diurnal lifestyle – Oilbirds and bats are nocturnal, while certain whales use their echolocation in the murky depths of the sea – so

---

*Opposite above: Some Swiftlets, like these Christmas Island Glossy Swiftlets (Collocalia esculenta natalis) form colonies in caves in total darkness. **Opposite below:** The eggs and young of a Black-nest Swiftlet (Aerodramus maximus). The nest is made mainly from saliva and feathers. Adults use echolocation to find the nest in murky caves.*

you could actually call swiftlets 'part-time' echolocators. You might also be surprised to know that the clicks they use are within our own audible spectrum. In contrast to the vast majority of bats, which carry out their business inaudibly as far as we are concerned, we can catch swiftlets in the act of echolocation, picking up long trills that somewhat resemble the sound of running fingers over a comb. Indeed, all of their clicks are below the 20 kHz limit of what is audible to us, yet they are fully sufficient for the swiftlets' purpose. As an aside, it demonstrates that people could, in theory, use this sense too, using echoes of their own voices to create an idea of where they are – some blind people have demonstrated this rudimentary ability.

And in case you are tempted to think that swiftlets are using somewhat rudimentary echolocation for themselves, consider what they are actually capable of doing. These birds both roost and nest in the dark, with zero light available, so at the very least they must be able to fly around without banging into the walls of their home caves and killing themselves. However, they are sociable and colonial, so swiftlets must also be capable of flying around without colliding with any of their colleagues doing the same thing. And thirdly, they must also find a suitable nest site, build a nest platform and raise their young in it, meaning that they need to know exactly where in the cave they actually are at any moment, and how to return to their site. This also suggests that these small birds have an excellent spatial memory.

Several puzzles excite those studying swiftlet echolocation. The talents described above are remarkable, and also baffling. Most swiftlets give off clicks in the frequency range between 1 and 10 kHz (well within our limits) which, by all accounts, should only enable them to resolve objects to a limit of some 34mm and no smaller. Yet field experiments have shown them to be able to negotiate past vertical rods of 10mm diameter and even lower (down to 6.3mm in one experiment). And while this should be possible using a faint ultrasonic component to their calls, physical experiments on swiftlet auditory neurons do not suggest that they can hear such high frequencies.

If swiftlet echolocation can resolve down to detecting such small objects, then why should the birds leave such an outstanding talent back at their home caves? They could be using it to catch food, just as bats do. Not surprisingly, those who study swiftlets, birds known to be highly active at dusk, have been looking for clues to nocturnal feeding for some time, but it clearly is not a staple strategy in the group. Several studies, including those of the Mountain Swiftlet (*A. hirundinaceus*), have proven that the birds only click at their caves. However, the Atiu Swiftlet, (*A. sawtelli*) and the Papuan Swiftlet, (*A. papuensis*) definitely click outside when foraging, presumably to obtain food. In itself that poses another question: why would some species use echolocation to catch aerial prey when others don't, the latter apparently missing a trick? Nobody knows.

One thing that they can detect unequivocally is a nest: the rate of clicking in every species increases when the birds approach their sites, consistent with the birds using echolocation to find them. The nests range from 50–100mm in diameter. Whether the birds find their precise site by detectable differences in the nest's structure, or whether they use the nearby topography of the cave or tunnel or building to home in on their own nest, is not known. Of course, swiftlet nests are astonishing in their own right. They are small platforms of hardened saliva, often in combination with other material, stuck to the vertical sides of the caves or buildings. The nest of the Black-nest Swiftlet (*A. maximus*) is made of saliva mixed with feathers, that of the Australian Swiftlet (*A. terraereginae*), as well as several other species, is saliva mixed with vegetation. Remarkably, the nest of the Edible-nest Swiftlet (*A. fuciphagus*), is virtually all saliva – and these nests are, as their name suggests, the main ingredient in genuine bird's nest soup, a multimillion dollar industry.

Therefore, the more you chase after a swiftlet, the more amazing its hidden life proves to be. Indeed, there is something else concealed, too. So far, nobody studying the physical attributes of swiftlets has found anything out of the ordinary in its auditory system, from the brain to the ear, nothing to suggest special abilities in echolocation or anything else. Presumably, one day they will find something, if they follow for long enough.

AUSTRALASIA

WHITE-WINGED CHOUGH

Our family group needs some extra help

White-winged Choughs (*Corcorax melanorhamphos*) are somewhat sinister birds. They shouldn't be even slightly alarming, because they are not especially big, no larger than a crow, and not predatory in appearance. They are also active in broad daylight in the brightly lit dry country of inland south-eastern Australia, and they cause no trouble, except to the insects that they dig from the surface of the soil or find under the leaf-litter. Yet they have something vaguely menacing about them. For one thing they are black, which is always a little unsettling, and they have peculiar bright red eyes and a long, curved bill. They also spend almost all day walking along the ground, which is very unusual in birds of that size, and they are unusually quiet, uttering only a few soft calls. That means that you can be peering at other birds in the treetops through your binoculars, and then suddenly you hear a rustle and a small party of White-winged Choughs almost materialises at your feet. It feels as though they – and it's always they – creep up on you.

As it happens, their terrestrial life is one more of hardship than menace. Theirs is an existence seemingly scratched out against the odds – which is, of course not so unusual among the human inhabitants here. Yet these are native birds on a truly ancient continent and they ought to be well adapted to their environment. Other birds and animals thrive here and yet, hidden away in the natural history of this unusual species are several indications that their lot in life is particularly tough.

White-winged Choughs live in social units of a handful to about 20 individuals, and the unit is strong and cohesive over a long period of time. Such a unit typically consists of a primary breeding pair and their various progeny produced over a number of seasons. On an average day all the flock members forage on the ground for food, moving along in measured fashion a few metres apart. Each clan has a territory that is vigorously defended from neighbouring groups. If two groups meet face to face the primary males engage first, and then it is all hands to the pump, like a saloon bar in a cowboy movie. Every flock member, from the breeding pair to the most recent offspring, contributes to territorial defence, at least during the breeding season.

This solidarity and co-operation extends to the annual breeding attempt. In contrast to a good many birds, even other group-living species, nest-building also engages everybody, from the oldest to the youngest, regardless of their degree of skill, and even if there are 20 members of the flock. The very

youngest do tend to take an observer's role, but second-year birds make a genuinely useful contribution. And they need to, because the nest of a White-winged Chough is high maintenance indeed. Most unusually, it is built almost entirely of mud and placed on a branch above ground. Such a rare structure cannot be rushed, because the birds must build it up layer by layer, and each layer must be caked dry and solid by the sun before another storey can be built. The result, a large cup, is impressive, if somewhat cramped.

Above: The only time you can see the white on a White-winged Chough is when it flies or ruffles its feathers. This bird lives mostly on the ground.

The nest might be a communal effort, but normally only the oldest female, the matriarch, actually lays eggs in it, 3–5 in all. Once the clutch is complete, though, the birds revert to their strategy of sharing out tasks, firstly incubation and then, to the very fullest extent, feeding the nestlings as they grow and fledge. The care lavished on these youngsters is exceptional: after leaving the nest at 18–20 days of age, the whole flock is at their beck and call for as much as 10 further weeks as they learn the vexed task of foraging. And even then, of course, they remain in the flock with their parents and siblings for the foreseeable future.

The first clue that not everything is easy for a White-winged Chough lies in the sheer effort given to the youngsters' welfare in the first place. But

Above: The peculiar nest of the White-winged Chough is like a large bowl made of mud.

another, more telling pointer is in the poor fruits of their toil. In one study undertaken in Canberra, no less than 65 per cent of nests raised a chick to fledging, but two-thirds of all the nestlings that hatched died in the nest, usually of starvation. Furthermore, the nesting success of a clan increases with its size, and there has never been a record of single pair succeeding without help, or even a group of three birds raising any young at all. It is obvious that White-winged Choughs find the burden of feeding young very difficult indeed.

Yet it isn't only feeding young that is the problem. White-winged Choughs also need to keep their territorial boundaries free from interference, and if they fail to do so, a breeding attempt can be impaired in a different way. Where a group is vulnerable by dint of its reduced membership, neighbours will be aware of this, and are not averse to launching attacks to destroy their rivals' exhaustively constructed nest. That way they beat down any competition for food in the vicinity. It's a brutal strategy, but an indication of the pressures these birds face.

One other way in which White-winged Choughs exhibit behaviour that appears to be on the edge of desperation is thought to be entirely unique to them. You will have noted from the above that nesting success is correlated, at least to some extent, to the size of the flock, and that the larger the group size the more young a clan can produce (really big groups sometimes actually attempt two broods in a season, one after the other). Well, to a White-winged Chough this is something of a headache, because the only legitimate way to increase the clan size is to produce more young, which, as we've seen, takes time and effort.

But perhaps there's a short cut to this? It seems that there is. And that, believe it or not, is to kidnap youngsters from another group. White-winged Choughs have been caught in the act doing exactly this. During conflicts with neighbours, fledglings, usually those just out of the nest and not yet able to fly, have been indeed snatched from a family group and kept within the group territory of the kidnappers. And although at least some of the affected youngsters have subsequently been seen to return to their home clan, others have apparently been co-opted into a clan that is not of their birth. That, indeed, is a desperate measure to increase breeding success. And it is also quite sinister.

FAIRYWRENS

What is the significance of the flower gift?

It's a touching scene. A male Superb Fairywren (*Malurus cyaneus*), in full iridescent blue plumage and in peak condition, brings a female a pink flower petal. The pink contrasts fetchingly with the lustrous cobalt-blue of his ear coverts, and as if to highlight this contrast he fans the ear-coverts so that he almost seems to be sporting a moustache. So intense is the colour of his plumage, and so perfectly executed is the performance, that an obvious impression is made on the female – enough for her to take the petal presented to her. Then, in what in human terms would be called a grand exit, the male immediately flies off on whirring wings, low through the bushes and out of sight, leaving the female clutching the petal.

It's a scene witnessed only occasionally, even by those scientists who have been studying fairywrens for years. These delightful birds, so well-known even to suburban Australians, are timid by nature and extremely skulking. It takes a great deal of patience, and plenty of ringing and trapping, to work out their complicated lives. However, what has been revealed is that scenes like the one described are not quite what they seem. Fairywrens have a way of doing things that is not yet fully explained.

Even the most casual of birdwatchers (or 'birdos' in Antipodean parlance) will immediately notice that fairywrens live in small groups that vary in number between five and 10 individuals. Studies have revealed that the flock consists of a pair of adults accompanied by their offspring of previous years, and this is typical of many group-living species. The accompanying birds are known as helpers, because they contribute to the rearing of the latest brood by bringing in food and performing other duties. The unit holds territory as a group, and all members of the flock take part in the defence of this territory. Members of the group are in each other's company all day long; they feed together, loaf together and roost together, often huddling together in physical contact and seemingly posing for the perfect family shot.

In the majority of group-living species it is only the adult female who breeds each year, and this is the case in Splendid Fairywrens (*Malurus splendens*), too. Very occasionally one of the younger females builds a nest and lays eggs, but she doesn't receive much help. Opportunities to breed are therefore limited. At any given time territories are fully occupied and only senior males and females breed, so in these comparatively long-lived songbirds most individuals have to wait several seasons before an opportunity comes along.

When opportunity does come, however, a particularly odd thing happens. More often than not, a sexual relationship happens not inside the cosy family unit, but outside it. In other words, if a male wishes to copulate with a female – and let's face it, he does – he will probably pay a visit outside his territory to do it. And a female, at the same time, seems to be predisposed to welcoming a stranger into her territory to inseminate her or, remarkably, she will undertake excursions on her own intitiative for the same purpose. The figures supporting this are enough to make eyebrows raise – in the Superb Fairywren, one study found that 76 per cent of all young are sired by males from outside the family unit.

Above: A male Superb Fairywren allopreens its mate. There is a genuine long-lasting bond between mates, but both sexes routinely copulate with other members of the opposite sex.

This figure is extraordinary. It isn't as if there is a problem between the male and female in a family unit, because they do produce young of their own. Neither does the figure get distorted by extra-pair copulations between junior ranks, because on the whole only the senior female copulates in any given unit, even if her suitors may be senior or junior males. No, the reason for the extraordinarily high extra-pair liaisons must be different.

Another piece of evidence suggests something unusual is happening, and that is the reaction of the incumbent males to strange visitors intruding upon their territory and inseminating their social mate. In most circumstances males would take exception to any kind of visit and violently eject a rival if he caught him in the act, so to speak, but in fairywrens, remarkably, incumbent males often don't intervene, even when they can see precisely what is happening. Their very passivity is extremely unusual, and puzzling.

Above: The Fairywrens are among the best known and most attractive of all Australian birds. This is a Splendid Fairywren.
Opposite: Extreme differences between the sexes, as in these Splendid Fairywrens, is often a sign of regular polygamy.

As yet, nobody has yet come up with a definitive explanation for the high rates of extra-group paternity. One of the most plausible is that it prevents incest. If a junior male inherits the dominant position in the group when his father dies, he does not then wish to be copulating with his mother or his sisters. However, the latter is unlikely because females usually move groups when young, and he is just as likely to meet them in next door's group as his own. Another possibility is that it simply allows females to increase the genetic diversity of their brood – but since this would be true of any species, that doesn't explain the extraordinary high degree of extra-pair liaisons.

As yet we don't know. But what we do know is that, in Superb Fairywrens, the wholesome scene described at the start is likely to be part of the birds' subplot. In early season, before nesting begins, males routinely visit females in their territories, very early in the morning. They come bearing petal gifts and depart with haste. Their message is clear: I'll be seeing you soon.

GREAT BOWERBIRD

Stage management by bowerbirds

The bowerbirds (Ptilogonatidae) are famous the world over for their elaborate constructions, 'bowers', which males build in order to impress females. The bowers have various different designs, but they all involve an extraordinary cost in effort and hours, and typically consist of thousands of different pieces. They are equivalent to some of the world's largest nests in terms of size and complexity, yet they have no function in bringing up young. They have no purpose at all, indeed, other than to draw in a female looking to mate. Remarkably, even then the bower itself is not enough to clinch a deal. The male bowerbird also has to display in his bower, and if his efforts are not up to scratch, he and his building work will be summarily rejected.

One can easily imagine that a population of exacting females and keen competitors has driven the evolution of bigger, better and ever more elaborate bowers. It seems likely that females will go for simple and obvious improvements to large and elaborate constructions. However, there is a point at which simply adding more material, making the bower bigger or putting in a greater variety of debris could reach its limit in the impression it has on females. And that isn't the only consideration. Since males routinely steal material from each other, there is also a limit in bower size that a male can reasonably defend. In such circumstances, perhaps there is a compulsion upon bowerbirds to build with guile rather than simply with brawn? Two recent studies in Australia suggest that this is exactly what they do.

The Great Bowerbird (*Chlamydera nuchalis*) is the builder of an Avenue Bower. This consists of two thick vertical walls of small sticks arranged parallel to each other so that there is a narrow passageway between them. The walls themselves are each 34–46cm tall and a good 15–20cm thick, so that the whole bower is about 40–60cm wide, and the floor of the bower is a pile of sticks raised a few centimetres above the ground. The passageway itself is about 50cm long. You might think that such an architectural marvel, which takes some 5,000 sticks to construct, would be enough to gain a female's attention. But no, the ground around one end of the Avenue is littered with large numbers of decorations, up to about 12,000 in all. These are usually snail shells, but many other items may also be gathered,

---

*Opposite: Another snail goes on the bower of a Great Bowerbird.
It may collect thousands of these, and place each one individually.*

including mammal bones, green fruits, leaves, bottle tops, sloughed lizard skins and a small number of red items. Even marble chips from nearby graves in cemeteries have been used. It is in the middle of these pieces of debris that the male Great Bowerbird performs his display.

Recently, researchers from Deakin University in Australia have looked at the precise arrangement of the objects placed around the Great Bowerbird's court. As in previous research they found that the bird invariably places smaller items close to the Avenue and larger items further away from it – indeed, the absolute size gradually increases with distance, so medium-sized snail shells or stones are placed in the middle between smaller and larger ones. The birds' behaviour suggests that this arrangement is important, because when objects were moved to interfere with the steady size increments, the birds noticed moved them back again. This made the scientists begin to look at the bowerbird's work from a different point of view.

When a female Great Bowerbird arrives at the bower, she takes up a position whereby it looks through the Avenue to watch the male displaying. When the scientists put themselves in the female's shoes, they realised that the steady gradation in the size of ornaments meant that the whole scene looks more regular than it would if there was a random arrangement of items, with objects both near and far appearing to be of more equal size to each other. This type of landscaping creates an altered perspective, perhaps making the whole scene look neater. Furthermore, within the regular pattern, the male himself might be more obvious.

But there is also a potential second advantage. By altering the perspective and putting the larger objects further away, the whole effect could also potentially make the male, displaying close at hand, appear larger than it actually is. And it seems reasonable to assume that, along with most of the world's bird population, females like larger males.

Do the male Great Bowerbirds do all this on purpose? Do they actually realise that their work gives forced perspective, the only known non-human instance of this trick familiar to human artists, architects and designers? So

Opposite above and below: The bower is finished and now comes the moment of truth – the female stands in mid-bower. Note how the male's pink crest is raised.

far, nobody can confirm that the birds know that this is what they are doing. Intriguingly, though, the trick only works when the female is standing in the right place, within the Avenue than the male has taken such pains to build. Do the males not only create the evened-out scene, but also ensure that the female stands in the right place to view from a forced perspective? It would be astonishing, and a source of wonder, if they did.

Meanwhile, another species of bowerbird, the Spotted Bowerbird (*Chlamydera maculata*) inhabits dry country to the south of the Great Bowerbird's range. This is a closely related species, and also builds an Avenue Bower with many decorations placed on the forest floor around it. The females of this species have an evident fondness for the colour and texture of the fruits of a shrub known as the Potato Bush (*Solanum ellipticum*), which are dark green, because there is a correlation between the number of fruits displayed on a bower with the owner's mating success.

Recently, a study at Exeter University, England, uncovered the fact that bowers occupied for a number of years have a higher proportion of Potato Bush plants around them than would be expected in a random distribution. Furthermore, when first occupying the territory and building a bower, the birds do not select locations rich in this plant. Instead, the area around newly constructed bowers held an average of 40 newly seeded plants a year. Overall the number of plants quadruples within 10m of the bower. Individual birds with more Potato Bush plants nearby typically have more of their seeds in their bower, leading to higher instances of successful mating for the males.

The male Bowerbirds don't plant the seeds, of course, but they don't eat them either, and by allowing them to shrivel on the ground, which is kept relatively uncluttered by other seeds by the birds, they allow them to germinate preferentially.

The 'gardening' actions of the male Spotted Bowerbird are the only known instance outside human activity where plants seem to be cultivated for reasons other than food. Whether intentional or not, the birds' favourite fruit flourishes as a result of their actions – and that seems pretty smart.

Opposite: There may be 12,000 items in the bower of a Great Bowerbird, and by no means do all of them need to be natural.

SOUTHERN CASSOWARY

Don't mess with this big bird

There is something primal in the fear of a big animal appearing out of
the forest. In some parts of the world, there is the genuine chance of being
ambushed by a dangerous carnivore, and elsewhere you wouldn't want to
spook large herbivores with enough bulk and belligerence to do you damage;
you are right to be wary. In Australia the dangerous animals are a lot smaller
and the patter of big feet is usually a kangaroo. But at least it does explain
why, when a Southern Cassowary (*Casuarius casuarius*) appears in front of
you, the natural reaction is a frisson of alarm.

The Southern Cassowary is a very big bird, the third largest in the
world behind the Ostrich (*Struthio camelus*) and the Emu (*Dromaius
novaehollandiae*). It stands up to a man's chest and, like those other two
birds, it has a long, reptilian-looking neck, which is bare-skinned and
pigmented electric blue. The face exudes attitude and not a shred of warmth,
and there is an outsize casque upon it that you can easily imagine could be
used for head-butting, and you'd be right (it might also be used as a shock-
absorber as the bird runs through the forest and hits an obstacle). The body
is massive and the legs extremely thick, as if the cassowary was built to
play rugby. The plumage is deep black, and rather hairy in appearance. It
allows the cassowary to melt away into the shady rainforests of northern
Queensland, disappearing from sight, despite its size, within a few footsteps.

And you really do want to see the cassowary disappearing, because it is
just about the only bird in the world that can do you serious harm.

The locals in Queensland will tell you that you could be killed. And it is
indeed true that there is a documented case of a young man dying in these
parts as a result of a cassowary kick. It was back in 1926, when two brothers
set out into the forest and found themselves facing the large bird. Unwisely
they tried to kill it, and the cassowary kicked at them in self-defence. One
brother was floored by the kick, recovered and ran off. The second brother,
a certain Phillip McClean, tried to club the cassowary but was also repelled
and fell to the ground. Tragically, the cassowary then kicked the 16 year old

---

*Opposite: This male Southern Cassowary is guarding its green eggs.
It is unwise to approach this large bird close to the nest.*

in the throat. Although the boy ran off, the wound proved fatal and he died shortly afterwards.

Although this is the only known Australian fatality, others have been reported from the island of New Guinea, not far to the north of the continent. Here the Southern Cassowary lives with two other, smaller but closely related species. Some of the early explorers to New Guinea recounted tales of villagers being killed by cassowaries, and there has been at least one recent rumour of an unprovoked attack that claimed the lives of two people. With birds and villagers living in close proximity over thousands of years, it does not seem too far-fetched to assume that cassowaries have, from time to time, fatally injured people. They have taken out dogs, too. There is a report that one even managed to kill a small horse. So how does this happen?

It turns out that the cassowary's main weapon is not its bulk, nor even the power of its forward kick. Instead it is armed. Each leg is supported by three toes, and on the middle toe is a razor-sharp claw up to 10cm long. This is the large bird's secret weapon, the part that does the damage. In theory, at least, it could disembowel an animal the size of a dog. In fact, it cannot inflict a wound of more than about 1.5cm in diameter. But if the puncture ruptures a major artery, then it is easy to imagine death from loss of blood. Furthermore, when they are kicking, cassowaries jump up and aim with both feet at the same time, doubling the amount of damage they can cause.

In recent years, cassowary problems have been on the rise in Australia, so much so that they have actually been catalogued. Up until 1999 there were 221 incidents around Cairns in Queensland, of which 150 were directed at people and 35 towards dogs. In addition to the single human death, there were six other serious injuries, including punctures to the flesh and broken bones. Cassowaries will jump on somebody who has fallen over, and continue kicking, and four of the most seriously injured people had been forced to the ground. In another recent incident a man fell into a lake trying to evade a cassowary, but was unharmed.

*Opposite above: Despite its enormous size (up to 1.8m tall), the forest-living cassowary easily disappears into the vegetation.
Opposite below: The thick legs and feet of the Southern Cassowary. Note the long, sharply pointed claw on each inner toe, a formidable weapon.*

Fearsome though they are, cassowaries are not accustomed to looking for trouble, notwithstanding the aforementioned figures. Indeed, any kind of attacks were very rare before 1985. Since that time, however, some individual cassowaries have overcome their fear of humankind, and have even wandered into towns. The birds are vegetarian, and some have become accustomed to being fed by people. Amusingly, these townie cassowaries sometimes kick out windows and doors, presumably becoming aggressive at the sight of their own reflections. The most recent attacks on humans have mainly been charges, as opposed to kicking, with head-butting. People out walking in the rainforest quite frequently recount being followed by cassowaries for minutes on end, the birds seemingly threatening rather than directly aggressive.

Tameness, it seems, does not suit these birds. Rubbing shoulders with us, apparently, increases their tendency to fight. Indeed, several of the most serious attacks on humans have been made by cassowaries kept in zoos. Furthermore, the recent claimed incidents of deaths in New Guinea are attributed to birds that had originally been brought up in human company and then released into the wild. Three-quarters of recent cassowary attacks have been made by birds soliciting for hand-outs of food.

There is something of a delicious irony in this. If you meet a cassowary, as I have, on a well-used trail where it is accustomed to people, you might find yourself at risk. If however, you are in the depths of the forest, and you come upon a very wild cassowary unexpectedly up close and terrifying, that is when, counterintuitively, you are probably safest in the large bird's presence.

Opposite: In Southern Cassowaries, the male is entirely responsible for incubating the eggs and looking after the young, which may take up to a year.

VARIED SITTELLA
Sittellas together

'Birds of a feather flock together'. It's a well-known proverb, and an indication as to how people often perceive birds – as social creatures, routinely gathering together in large numbers. There is an implication of benevolence in the proverb, and it is often used as an idiom alluding to genial social behaviour among people.

As most enthusiasts know, bird flocks are really a double-edged sword, full of rivalry and disputes. In the majority of flocks, especially those that persist over a long period of time, a clear hierarchy develops in which certain birds are dominant over others, for feeding sites, nest sites and roost sites. Stepping out of line at best produces a sharp physical response from a dominant bird, often in the form of a supplanting attack, when one bird displaces another physically from a perch. At worst, birds can be killed. This sometimes happens in flocks of Ruddy Turnstones (*Arenaria interpres*), when individuals feeding in 'unauthorised' locations are physically attacked.

Being at the bottom of a dominance hierarchy is a harsh station in life, and its consequences can be brutal. At the feeding station, subordinate birds may be kept from acquiring the food they need; over time they lose body condition and can die as an indirect result of their place in society. In the breeding season subordinates can be barred from territories or suitable nest sites. And, in the night roost, a bad position can be fatal. Individuals on the edges of communal roosts are likely to suffer higher mortality, both from predation and from exposure to the cold.

This somewhat bleak picture of the life of a subordinate bird is, regrettably, quite a common one. But it isn't entirely universal. A most refreshing exception was recently uncovered in a terrific study by R.A. Noske in New South Wales, Australia, and the subject was a widespread inhabitant of light woodland and bush, the Varied Sittella (*Daphoenositta chrysoptera*).

Sittellas are small, restless nuthatch-like birds that live in small groups averaging five individuals. In common with many such group-living species, the unit is based around an adult pair, with various hangers-on, including a selection of the previous year's youngsters, which sometimes swells the group numbers well into double figures. During the day the flock lives in close proximity, feeding in the same trees at the same time, moving from place to place within sight of each other, and regularly settling down for bouts of mutual preening. There is a hierarchy, though; the senior adults determine

when and where the group will forage next, and they also determine when the group sleeps during the night. However, this is a hierarchy like no other, and its unusual nature is reflected in the sittella's roosting arrangements.

As the sun comes down on another busy day of foraging, the dominant male in each sittella group selects the nightly roosting spot. It is typically a very high tree fork between nine and 16 metres above ground; the birds use the lower branch of the fork, such that the innermost sleeping bird is wedged in, and the other part of the fork acts as a roof over the huddle's head. Once it has selected somewhere suitable – a place that may be used for many nights in a row – the dominant male makes known to the rest of the group that plans have been made. Nobody rushes to the scene; instead the appearance of the other members of the flock is staggered, with each individual coming in at intervals of 30–60 seconds, so as not to attract attention from predators.

Above: The Varied Sittella spends much of its time foraging on the surface of tree trunks and branches, in a similar manner to a Nuthatch (Sittidae).

First on the scene are the subordinate, non-breeding males. Observations made on sittella flocks indicate that this individual takes the innermost position, wedged at the junction of the fork. Meanwhile, the dominant, breeding male huddles next to it, and the rest of the huddle will form, not by birds filling in positions on the inside, but in fact squeezing in between settled individuals. The breeding female is usually next in, but is sometimes beaten to it by a junior male. Either way, the last individuals in are always the immature birds that are only a few months out of the nest. These younger birds always take the innermost positions in the huddle. Once the huddle has settled down, the dominant bird in the group, the breeding male, is always on the outside. Sometimes, when the group is large and the roosting branch is short, he will actually topple off the end of the perch as his subordinates settle in.

Believe it or not, the sittella parties don't just have a regimented order of fitting into the roosting huddle, they also have consistently different sleeping times. You can tell when a sittella is asleep because when it first enters the huddle it employs a special posture, with head down and tail up, whereas when slumbering it is more horizontal. By watching carefully, Noske discovered that the youngest birds fall asleep first, while the last bird to do so is the dominant male. The time interval is considerable: the last bird may finally fall asleep 48 minutes after a youngster.

*Above: Living in extended family groups, sittellas are extremely sociable birds carrying out most activities communally. This group is taking a break to preen. **Opposite:** At roosting time, subordinates invariably take the innermost position in a sittella huddle.*

Since the cold is not normally an issue for sittellas in Australia, it is clear that the nature of the huddling both reinforces social bonds and provides the flock with extra security from predators. It is quite clear that the dominant male stays awake in the darkness to ensure that the coast is clear. So, not only does he take the outer position in the huddle, but also keeps a lookout for the benefit of his flock.

If you want an idea of just how unusual this delightful arrangement is, you need look no further than the case of another well studied species, the Eurasian Long-tailed Tit (*Aegithalos caudatus*). These birds live in similar family groups and also huddle at night. When it is cold, physical huddling effectively makes them into a single larger organism with a smaller surface-to-volume ratio, reducing the heat loss of each bird. However, birds on the outside of the huddle are more prone to heat loss, as well as predation. In this case the dominant birds perch themselves unashamedly in the middle of the huddle. Even as parents, they let their own offspring buffer them from the cold. It's a far cry from the sittella's sacrificial care.

NORTH AMERICA

WHITE-THROATED SPARROW

A tale of two sparrows

Consider the following description comparing two species of sparrows (Emberizidae). Maybe you can guess what they are? Both are widespread in North America. One of our sparrows prefers open areas, including gardens and scrub and woodland edge. The other species thrives in much denser vegetation, including undergrowth, and has a slightly lower-pitched song of longer wavelength that passes better through this sort of habitat. One species is an indefatigable singer and, unusually for a small bird, both sexes sing frequently. Male and female both indulge in intense territorial skirmishes and aggressive displays, which suggests that the females are well in touch with their masculine side. The other species, on the other hand, sings very infrequently, although the male's song output does have a tendency to increase as the season progresses. However, it is a shy performer. Compared to the first species, it is remarkably placid and rarely gets involved in any kind of skirmishes, even with its nearest neighbours.

The breeding strategies of the two species are similar, but have quite different fringe benefits. Both follow the typical small bird arrangement known as social monogamy, in which a male and female team up and divide roles between them in order to carry out a breeding attempt. However, males of the first species are prone to cuckolding their neighbours, readily and routinely mating outside the pair bond, despite their cosy domestic arrangements. Males of the other species, on the other hand, do not do this. The difference in character is such that males of the second species are family-oriented, devoting a great deal more time than their sister-species to feeding the young. The males are far more paternally inclined than their testosterone-fuelled neighbours.

Seasoned North American birders will probably have twigged the point of this comparison, but as a final clue, let me outline the physical differences between them. These sparrows are very similar, but in one species the breast is plain slate-grey, while in the other there is usually some brown streaking. And while one species has an immaculate white eyebrow contrasting smartly with the black eye-stripe and crown-stripe, the other has a brownish supercilium that swamps much contrast with the duller black stripes.

Opposite: The immaculate, clean-patterned and assertive White-striped version of the White-throated Sparrow sings almost all the famous 'Old Sam Peabody' songs beloved of North American birders.

By now you should have picked up that these two species exhibit profoundly different behaviour. So which ones are they? Step forward, the White-striped (WS) version of the White-throated Sparrow (*Zonotrichia albicollis*), and the Tan-striped (TS) version of the White-throated Sparrow. They are not two species at all, but two colour morphs of exactly the same species. The White-throated Sparrow is one of the continent's best known and best-loved species, its delightful, cheery and very clear 'Old Sam Peabody, Peabody' song making it instantly recognisable. Intriguingly, the status of this bird as one of America's top songsters is mainly down to the efforts of WS individuals.

WS and TS versions occur wherever the White-throated Sparrow occurs, so every population contains chalk and cheese. There are no intermediates. In the winter the forms are hardly distinguishable, but their differences magnify during the breeding season. The whole White-throated Sparrow population is split roughly 50:50. The key to maintaining the divide is that one morph always prefers to breed with the other morph, its opposite. Aggressive WS males have the benefit of gentler TS females, while the paternally inclined TS males find company with high-spirited WS females. This is known, rather splendidly, as disassortative mating.

The case of the dimorphic sparrows is unique in the world of birds, especially because the differences are expressions of a stable genetic quirk. The second pair of chromosomes in WS and TS birds are different, with one pair of the same shape in Tan-striped individuals, but a dissimilar pair in White-striped birds in which one of the chromosomes is partially inverted. It seems that the inversion contains a whole block of genetic material, a super-gene, responsible for expressing the differences. The arrangement apparently represses recombination and, together with the fact that birds almost always pair up with their opposites, this ensures that the dimorphism passes down the generations.

There is no other bird in which there is polymorphism of this particular nature, and the White-throated Sparrow has become a poster bird for geneticists wishing to study the links between genes, hormones and life history. There has been much promising research. Already scientists have measured a clear link between the testosterone levels in the males of the two morphs, with the WS not surprisingly showing higher levels. These individuals also have larger testes and cloacal openings.

Meanwhile, TS females have higher levels of oestradiol than do WS females, which is equally predictable. Before the breeding season begins, WS males show enhanced corticosteroid levels compared to TS males, so if you wished to describe a WS male as like a TS male on steroids, you wouldn't be too far wrong.

For the moment research continues in this area, but each new study seems to discover more ways in which WS and TS birds differ, despite them being variations of the same species.

And perhaps the most fascinating discovery of all has been the long-lasting nature of the difference: gene flow between the divergent regions of the second chromosome is estimated to have ceased about 2.2 million years ago. In that case, the chromosomal differences occurred before the White-throated Sparrow diverged as a species in its own right. The polymorphism pre-dates the speciation.

Above: The quiet, hen-pecked behaviour of male Tan-striped morph White-throated Sparrows is so different to their White-striped colleagues that they could almost be different species.

BLACK-CAPPED CHICKADEE
Memories of garden birds

When it comes to comparing the abilities of birds and people, it's easy to accept there are some ways in which our fellow creatures outperform us. We cannot fly, for one thing, and we are not as good as birds are at travelling accurately for a long time in a particular direction, while birds routinely migrate vast distances without any kind of help. We find it easy enough to appreciate that birds are superior in the sensory department: they see more crisply, have more of an ultraviolet component to their vision and some, such as owls, also have extraordinary powers of hearing. As it happens they also have a magnetic sensitivity that we probably lack. Birds are endowed with many physical gifts.

However, when you sit down to have coffee in the kitchen and idly watch the comings and goings of the birds at your feeders, would you be surprised to know that some of your garden companions might actually out-perform you in a more cerebral function? A variety of studies on the cognitive abilities of tits and chickadees suggest that this is indeed the case. It seems that some bird table residents have, compared to us, a superior spatial memory.

Species such as Black-capped Chickadees (*Poecile atricapillus*) are mainly vegetarian in the winter, subsisting primarily on seeds and nuts. There are two characteristics of these plant products that make them unusual in respect to other kinds of food. For one thing, they do not spoil quickly. Many have hard cases that are adapted to protect the seed inside from decomposing, giving them the longest possible time to be discovered by a dispersing animal. Another ecologically important aspect of seeds is that they are produced in vast quantities at only one season of the year, at least in the colder, temperate parts of the world, and that is during the autumn. The first characteristic makes it possible to store them, and the second characteristic makes it highly advantageous to store them.

In the autumn, therefore, and also during the winter when they are foraging from artificial feeding stations, Black-capped Chickadees don't necessarily eat everything they find. Instead they will carry food items away and hide them in all kinds of different places, including clumps of lichen, moss, soft ground (even snow), dry leaves, crevices in tree trunks and behind loose bark. During the summer they will also store some animal matter, including dead spiders, although only for brief periods, but overwhelmingly it is nuts and seeds that are secreted away. These items are, remarkably, stored singly, just one seed per hiding place. It would be disastrous for a bird to put

many items into a single cache, if that site was then discovered and emptied, ruining hours of effort. By scatter-hoarding in this way, the chickadee insures itself against theft.

There are two reasons why an individual might store food away. The first is to cement its claim to the food items. The bird feeder is a busy place, and it might well pay an individual to pick up a seed and take it right away from the public arena, so to speak, so that it can eat it later, when things have died down and it is less likely to be disturbed or robbed. However, another reason why chickadees might cache food is to build up reserves of food that they can tap into at a later date. This could be invaluable as winter progresses and the external seed supply is expected to diminish (it doesn't at bird feeders, of

Above: A Black-capped Chickadee collects a seed. It stores thousands of small items like this away, to retrieve later when needed.

course, although it would in the natural environment). If an individual has a hoard of food hidden away, this should buffer the owner against any severe conditions that make daily feeding difficult or impossible.

However, a food cache is only any use to a bird if it can retrieved later, and that requires certain conditions to be met. Firstly, the scatter-hoarding bird must be resident in the same territory over a long period of time; it cannot be wandering around the woods, far from its caches, and it does not want trespassers in its territory either. Black-capped Chickadees meet this condition, being long-term and often lifelong residents in the same piece of wood or backyard. The second condition has profound implications. If scatter hoarding is to be of any value, the bird must be able to remember where it has put its bounty.

In fact, there are various ways in which a bird could give itself a chance at recovering the caches. It could simply search in places where there is a high chance that it will run into them again, randomly, or use a particular searching method that should lead it to the right places. It could even mark the cache sites. However, experiments on captive Black-capped Chickadees in the laboratory have shown that the birds do work from memory. For example, scientist David Sherry and his co-workers provided several chickadees with potential storing sites in 70 holes drilled in an aviary. Once they had lodged four or five seeds away the birds were removed, the aviary was cleaned, the seeds were all taken away and the handful of storage sites were covered by flaps, making them look different. Despite this, when the chickadees were put back in the aviary, they spent 10 times longer looking in the sites they had used previously compared to ones they had never used. This strongly suggests that they were working from spatial memory.

Several further studies have revealed even more extraordinary details about the chickadees' feats of memory. It turns out that, not only do they remember their cache sites, they also remember the locations of any seeds or nuts that they encounter while foraging but do not pick up. Astonishingly, they only use a cache-site once, and they can remember the sites for at least a month, and almost certainly for much longer.

All of this would not be so remarkable if chickadees only used a few sites, but they don't. They use a lot, depending on where they live – birds breeding in the far north of the continent use more than their southern relatives. They

definitely use hundreds, because they may store 100 or more items in a single day. They might use thousands – a close relative in Eurasia, the Willow Tit (*Poecile montanus*), is thought to store away half a million seeds a year. This species apparently remembers where 90 per cent of its caches are stored, and the Black-capped Chickadee is likely to approach this.

How do the birds achieve such feats of memory? It just so happens that their hippocampus, the part of the brain associated in humans with spatial memory, is comparatively larger than it is for similar species that do not store food.

Therefore, if you ever find yourself in that annoying everyday situation where you have misplaced a letter, or some item of stationary, or your car keys or some other necessary item, look out of the window at the bird feeders. Birds there, including Black-capped Chickadees in North America, may well be better able to find what they need than you can.

Above: Black-capped Chickadees are habitual visitors to backyard feeding stations.

CLIFF SWALLOW
Unnatural selection

The Cliff Swallow of North America really isn't a cliff swallow any more. When it was first named and described in 1817 it used to build its nest only on vertical rock faces, such as cliffs, rocky outcrops, gorges and natural walls. The nest, which is shaped like a gourd with a narrowing tunnel entrance at the top, is perfect for such sites. It is made up from hundreds of pellets of dried mud, and can thus be stuck on to any suitable surface. Placed just below an overhang for shelter, it can be judiciously constructed at any height above ground, well out of the reach of predators.

These days, however, only a small minority of Cliff Swallows (*Petrochelidon pyrrhonota*) breed in natural locations. Now they prefer buildings, bridges, culverts and other artefacts, which have proliferated all over the continent as humans have spread to every corner. The advent of man has seen a boom in Cliff Swallow numbers and range. Once Cliff Swallows were confined to the Great Plains and westward, but in the last 150 years they have conquered almost the whole of the eastern side of North America as well. As people have been building, so have the numbers of Cliff Swallows.

You might thus be able to say that, in some ways, the Cliff Swallow has been shaped by its association with people. Certainly this applies to its status and abundance. But a recent long-term study in Nebraska suggests that the Cliff Swallow has also been shaped by its relationship with us in another way, too – physically.

This remarkable and perhaps unprecedented finding has come about in a deliciously serendipitous manner. A husband and wife team from the Universities of Tulsa, Oklahoma and Nebraska-Lincoln respectively, have been studying colonies of Cliff Swallows for more than thirty years. Their study site is the insalubrious roadsides of western Nebraska, where the birds are abundant in bridges and overpasses and other road-building artefacts. Charles Brown and Mary Bomberger Brown have been focusing their attention on all kinds of aspects of Cliff Swallow social behaviour, work that involves trapping and colour-marking hundreds of birds from the different colonies and taking a variety of measurements. In the course of their studies they happened to collect specimens of any dead birds that they found; those killed by traffic, those killed by happenstance, and some that had unfortunately perished in the nets. Little did they know at the time that these specimens would provide them with some of their most intriguing and extraordinary results.

During the course of their summer work (the birds migrate to South America in the winter), the Browns did notice that, over the years, the number of dead birds that they found by the roadside dropped. While they picked up some 20 road-killed swallows in the course of the season in 1984 and 1985, just after they began their studies in 1982, in the most recent years the numbers had dropped to just five a year. What made this observation so intriguing – and they might never had noticed it otherwise – was that, at the same time that roadkill numbers were declining, the overall population in the colonies had doubled. It appeared that something was causing the number of collisions between bird and vehicle to decline. The birds were apparently better at avoiding cars.

Above: Cliff Swallows collect mud for their nests. They often do this close to roadsides, and have adapted to dodging passing cars.

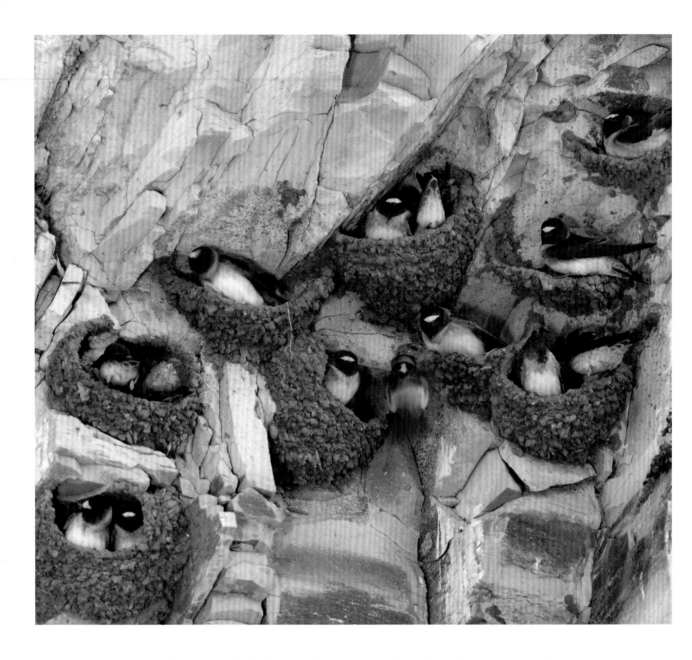

The Browns checked various factors to see what else might have changed, including the amount of traffic. However, this had remained much the same, both in volume and in patterns, and therefore was unlikely to account for the difference. They checked predation and disease, and looked into whether the population or habits of scavengers had changed in a way to affect the number of swallow bodies left around to be picked up by people. No clear changes in any of these variables were apparent. But what they did discover, completely to their surprise, was that the Cliff Swallows themselves had altered.

The Browns had two perfect data sets to compare: individuals that had been killed by cars (104 individuals), and those that had been killed in other ways, and thus could be considered as representative of the rest of the population (134). When they measured the length of the wing of each set, they found a mean length of 112mm for the road-killed swallows, and 106mm for the rest of the casualties. In other words, the individuals that succumbed to collisions with vehicles had the longer wings.

A still more telling statistic emerged from comparing the wing lengths of the complete data set of living birds measured over the years. The Browns found that the overall mean wing length in their colonies had declined from 111mm in 1982 to 106mm in 2012. And while that is a mere 5mm per wing, it constitutes a 1cm average reduction in wing-span over 30 years.

If you analyse the two sets of data, it is hard to escape two conclusions. The first conclusion is that longer-winged birds are more likely to be killed in road collisions compared to the rest of the population of Cliff Swallows; and secondly, over the years, natural selection has favoured the shorter-winged individual. Shorter-winged birds are more likely to survive and pass on their tendency for reduced wing-span to their offspring.

What advantage might a shorter wingspan bring? The Browns contend that shorter wings make the birds more manoeuvrable, so that they are able to perform sharp turns more effectively and thus dodge traffic more successfully. Separate studies on flight dynamics concur with this, and also show that birds with shorter wings can take off from a surface, such as a roadside, more easily.

It is not possible to prove without doubt that the Cliff Swallows of Nebraska are evolving shorter wings to avoid traffic. There could be any number of other factors at play. However, if nothing else it does prove that the birds at least have the capacity to adapt quickly to anthropogenic factors.

And after all, the Cliff Swallow has form when it comes to reinventing itself. It has already benefited hugely to adapting to nesting on structures built by people. Why shouldn't it adapt to moving objects as well?

Opposite: Cliff Swallows are very sociable birds, and some colonies may contain 3500 active nests. Apart from cliffs, they often select buildings and bridges as nest sites.

HARRIS'S HAWK

The hunter-gatherer

If small animals had nightmares, Harris's Hawks (*Parabuteo unicinctus*) would surely loom large. Many diminutive birds and mammals endure daily the constant threat of strikes by a bird of prey, such as a hawk. They are compelled to counter this with unrelenting vigilance, in which the briefest moment of lessened intensity could prove fatal. You might say that predation by raptors makes the quarry what it is: fast-reacting, fast-fleeing, never off guard. With most raptors the threat is deadly, but at least the danger comes in a factor of one. With Harris's Hawks, the terror is multiplied. These are the only raptors in the world in which the hunters form groups to gang up on their prey, with anywhere between two and six talented and relentless individuals making up their hunting parties.

Co-operation in predatory birds is comparatively widespread, but the largest killing unit is usually a pair. In the case of Golden Eagles (*Aquila chrysaetos*), for example, one partner flushes prey while the other intercepts it as it tries to escape from the first bird. It is easy to see the advantage in co-operation between a pair, because it pays them to act in each other's interest. In the case of Harris's Hawk, the unit is still a family, or breeding group, but these birds take the sophistication to a level unique among any kind of land-based avian hunters. They work as a well-drilled team.

Before a hunt, the group gathers at a favoured location, so all the birds may perch in the same tree, or even the same overhead wire. This is known as an Assembly Ceremony. The birds give the impression that they are discussing strategy, or that they are like a basketball team psyching themselves up before a game. At some invisible prompt they set off as one into the desert. Their initial searching style is for each bird to perch well above ground, scan for prey for a short time, just five minutes or so, and then to fly onwards for 60–300m on to a new perch. The birds may just carry on like this for hours, keeping the other members of the party in sight. It is quite usual for some individuals to duck out after a while, only to return later on.

Once a member spots prey, it alerts the others and the hunt may progress along several different paths, depending upon the situation. If the prey animal – a rabbit, perhaps, or a quail – is out in the open, it may provoke a relatively instant co-ordinated attack. This is the simplest form of capture, although still truly compelling to witness, in which each bird converges upon the prey from different directions at roughly the same time. All escape routes are cut off before the unfortunate quarry is able to dash for cover.

Sometimes a prey animal does manage to take refuge within nearby dense vegetation, and if it is a small hole-dwelling mammal it will be out of reach. If, however, it is simply trapped, the Harris's Hawks have a strategy for turning it into a meal. One or two members of the party enter the vegetation with a view to flushing it out, while the other members scan the scene from high perches and, if possible, grab it when it makes the break. They often fail, of course, but this is the sort of situation where a hunter on its own has little or no chance of success.

Above: Harris's Hawks hunt together, and may have to defend their share of the prey after a capture.

Sometimes a group of Harris's Hawks will come upon a particularly high value prey item, usually a rabbit such as a Desert Cottontail (*Syvilagus auduboni*) or a Black-tailed Jackrabbit (*Lepus californicus*). These fleet-footed animals are particularly difficult to catch, and pinning them down involves yet another form of co-operative hunting. In this case, the birds simply chase the victim for much longer than they normally would, wearing it down by taking turns to be the lead bird in the chase, working like a relay team. Responsibility passes from one member to the other to keep the animal in sight, and keep it moving, until an opportunity arises to go in for the kill – if,

Above: Larger mammals such as this rabbit are high-level captures for Harris's Hawks, almost impossible to pull off without a co-operative effort.

for example, the rabbit finds itself well out in the open, or it makes a wrong move bringing it within range. Even within this relay hunting, two birds may co-operate within the same move, one flushing and the other trying to intercept the fleeing animal.

Reading this, it might well have occurred to you that, in these co-operative hunts, the numbers of birds taking part matters. One might expect that the capture success rate should increase in parallel with the size of party. Not surprisingly, this does seem to be what happens. Indeed, in one study in Arizona, it was found that birds perch-hunting on their own achieved a capture rate of 20 per cent of each attack producing any kind of prey, while two birds caught 32 per cent, three had a success rate of 40 per cent, four of 38 per cent and five of 50 per cent. The figures are even starker for the captures of various species of rabbit. A study in New Mexico revealed that, unless there were at least four members in the party, they couldn't catch any rabbits at all. In time units of 50 hours, parties of two to three birds did catch some food, but not these high-value targets. If there were four birds in the party, they caught, on average, 1.7 rabbits every 50 hours, while if there were six birds, they caught 3.9. From this last figure alone, you can immediately see why Harris's Hawks are so keen to form their hunter gatherings.

Of course, there is a downside to foraging in a group, and that is the sharing of the spoils: the more members of the group, the smaller share each will get of the food. And Harris's Hawks do share spoils, make no mistake, even though there is a dominance hierarchy within every group. Co-operative hunting would soon break down if individuals hogged food, or cheated, because it would pay individuals to hunt alone instead. In the New Mexico study mentioned above, the researchers concluded that parties of five were best in terms of the value to each bird of the food gathered.

The Harris's Hawk is certainly a remarkable species, but perhaps the most astonishing aspect of its co-operative behaviour is that it is not universal throughout this species' wide range. Even in Texas, next door to New Mexico where it routinely hunts in groups, very few observations of co-operation have been made. Neither in Chile, where it has been well studied, nor anywhere else in South America has the Harris's Hawk been seen hunting in packs. Bearing in mind that this hawk is a more sophisticated communal hunter than any other bird in the world, you might think that it would use it given any opportunity at all. What a talent going to waste!

MARBLED MURRELET

Breeding in a different world

Take a flight of fancy for a moment. Imagine that you are following a bird in your backyard from its foraging site to its nest, taking a caterpillar to a brood of chicks – your bird could be a warbler, or chickadee or tit. Envisage that there is a remote camera on the bird's neck and you can see wherever it goes, from its own point of view. You will begin in the claustrophobic foliage of the canopy of a tree, and then something weird is going to happen. You are expecting the bird to descend down to a crevice in a tree, or a patch of dense foliage where its nest is located. Instead it takes off up into the sky and heads off out of your neighbourhood. To your astonishment it settles into a journey way out of familiar territory, flying a good 100–200m above ground. The kilometres tick by and eventually the coast appears in your sights, and then you are flying over the sea; what on earth is this small, insectivorous bird doing? The answer comes as, eventually, you lose height and finally land on an offshore oil-rig. There, in the wall of the structure is a small hole. The caterpillar in your bill is delivered, and you begin your bizarre return journey back to the foraging site.

Absurd, isn't it? And probably no bird has ever performed such a journey. It seems entirely counterintuitive – biological and ecological madness, even a bit silly. But there is a point to this story, because it demonstrates the bizarre and unexpected behaviour of one of North America's least-known birds. The nest of this species wasn't discovered for the first time until 1974, after more than 100 years marked by unsuccessful searching. And no wonder.

The bird concerned is the Marbled Murrelet (*Brachyramphus marmoratus*), a relatively common seabird found all down the Pacific Coast. It is a member of the auk family (Alcidae), which contains well-known species such as puffins, guillemots and auklets, as well as other murrelets. These birds are familiar to birders mainly because they have a habit of forming large, noisy colonies on tall sea cliffs and offshore islands, in sites free from ground predators. They typically commute from their nest sites to the open sea, where they forage. For many years, ornithologists assumed that they would find Marbled Murrelet colonies alongside other auks, and that the missing birds would eventually conform to the family pattern. As the years passed, people assumed that they simply hadn't found the right stretch of coast, or the right island. There is a lot of ground to cover between California and Alaska.

Marbled Murrelets look like, and are, quintessential seabirds. They spend most of their time swimming in the ocean, where they can ride out storms

and dive under the surface of the water to catch small schooling fish, such as anchovy and herring, as well as some crustaceans. Being small, they have an unsurprising tendency to occur in more sheltered waters, such as bays, coves and fjords, typically within 5km of the shore. But in this way they differ very little from many of their close relatives, and there is nothing physical to suggest that Marbled Murrelets should do anything bizarre and unexpected.

But time passed and it became clear that the mystery auk – 'The Enigma of the Pacific' – had a trick up its sleeve. Where on earth could its nest-site be – somewhere up in the trees?

Well, yes, as a matter of fact. That's exactly where the Marbled Murrelet does nest – in the middle to top third of the crowns of tall, old-growth forest trees. A small percentage of Alaskan birds build instead among crevices in rocky slopes on the ground, but the remarkable arboreal site is the norm, among the warblers and thrushes of coniferous forest, as far removed from marine cliffs and islands as you could possibly imagine. And still more incredibly, these canopy sites can be well over 50km from the coast; the birds

Above: A typical birder's view of a Marbled Murrelet gives no clue as to this bird's unusual life in the forest treetops.

travel to and from them under cover of darkness (or the long dusk of the Arctic night). Another difference from the family trend – hardly unexpected in the circumstances – is that pairs nest singly, not in colonies. But they prospect for sites and almost certainly compete for the best ones, as would any other seabird.

The precise demographic of the site is a horizontal platform, such as a broken-off branch, pile of needles or growth of creeper. The birds don't make any nest as such, but nestle the single egg into moss or some other soft substrate; some have been found on top of bird or squirrel nests. In California, a few sites have been found where the egg has simply been laid on a bare branch, but bearing in mind that most nest sites are 9–12m above the ground, that would seem to be risky at best. The auks' favourite trees are Douglas Fir (*Pseudotsuga menziesii*), Alaskan Yellow Cedar (*Chamaecyparis nootkatensis*) and Western Hemlock (*Tsuga heterophylla*). All of these are majestic conifers with dense foliage, and they are often covered with copious moss and lichen. When you walk in these forests there is such a mass of foliage above head height that it is perfectly understandable that the Marbled Murrelet's nest site remained undetected for so long – along with the sheer unlikelihood that a seabird would commute so far deep into the hinterland.

Not surprisingly, nesting in the tree tops makes some of the breeding duties of Marbled Murrelets tricky. Spare a thought for the single chick, the fledgling seabird that hatches in a treetop. Its first independent task in life is to make sure that it has exercised its wings enough to undertake a very long and demanding maiden flight. And even before it sets off, it has to pluck up the courage to launch into the air over its alien world.

And what about the practicalities of feeding the chick? The parents must find shoaling fish in the ocean, dive down to catch them, set off inland during the early morning or evening, make a long flight and find their way to the nest in the dark. It isn't quite the flight of fancy described at the start of this chapter, but it isn't far off.

Opposite: A rare glimpse into the early life of a Marbled Murrelet chick, high up on the branch of a conifer.

SOUTH AMERICA

ANDEAN COCK-OF-THE-ROCK

Working together with its friends

You don't have to delve too deeply into bird biology before realising that almost all aspects of a bird's life are deeply competitive. The struggle to find food, for example, has clear winners and losers. Rival birds seem to fight over just about everything to do with breeding: territories, broadcast airwaves, nest material, mates. Where everything is in short supply it is hardly surprising that altruism is as well.

The accepted wisdom does, however, come under strain when you consider the breeding behaviour of the Andean Cock-of-the-rock (*Rupicola peruviana*). Andean cocks-of-the-rock have the same needs as other birds, and are just as competitive. They spend a lot of time fighting. But Andean Cocks-of-the-rock, when they are displaying to females, have a male partner. The partner helps them to display effectively and a given individual cannot successfully court a female without its partner's help. Yet it's unclear how, if at all, these colleagues benefit in exchange for their efforts.

Cocks-of-the-rock belong to the cotinga family (Cotingidae) which, even by tropical bird standards, includes some astonishingly colourful and bizarre forms. Long ago the ornithologist David Snow worked out that species of birds subsisting on an entirely predictable diet, such as fruit, that was unfailingly available all year and easy to find, have the 'spare time', so to speak, to devote themselves to evolving wondrous displays and opulent plumage. Cocks-of-the-rock eat fruit. Thus the males do indeed display for hours every day, usually in the morning from sunrise until perhaps nine o'clock (coffee, shall we say?) and then again, more listlessly, in the afternoon. They gather what they need to eat in short bursts and the next day the same thing happens, and for the next days and weeks and months. They are emancipated from those time-consuming jobs such as nest-building and feeding young (the female does all of these on her own), and thus they spend a considerable proportion of their lives on the display ground, giving themselves over to impressing every female that deigns to visit.

If you visit the humid montane forests where these birds occur, and settle yourself in front of a display-ground, or lek, it will soon dawn upon you that performing cocks-of-the-rock spend their hours almost exclusively in

Opposite: The Andean Cock-of-the-rock is a member of the Cotingidae, a bird family renowned for their bright colours and elaborate displays.

male company. Peering through into the dappled shade where the display perches are, in the mid-canopy of the forest, all the action and kerfuffle comes from these gaudily-clad birds. They are medium-sized, equating to a Blue Jay (*Cyanocitta cristata*) or Eurasian Jay (*Garrulus glandarius*), perhaps, and are essentially black-and-crimson, although your eyes are drawn only to the pure, intense crimson, which seems to glow,, even in the shade. This colour covers the breast and back and all of the head, while the wings are tail are black, and some of the tertial flight feathers are light grey. The head bears a rounded crimson crest, which lifts from the nape and fans round to cover the top of the bill, and the eye is pale, contrasting with the crimson and somewhat resembling those false eyes you see stuck on to soft toys. The performers' appearance is dramatic enough, but as soon as it is light the birds start to bow forward and to jump excitedly up and down on their perches, and utter a truly extraordinary series of grunts and squeals that is decidedly pig-like, although drier and not as deep. The male cocks-of-the-rock also frequently move perches, so that the comings and goings on the lek are rapid and energetic. The backdrop of often quite dense foliage, with each branch covered with over-exuberant epiphytes and lianas, makes it hard to follow what is going on, but patient observations have gradually begun to make the picture clear.

The lek is the display ground for anything up to 15 males, although there are usually many fewer. Despite the comings and goings, every male has his own 'court' and holds this space long-term, save for the occasional changes caused by successful intrusions by a new or younger male. The males are within earshot of each other – their calls can be heard from 100m or more away – but they are not especially close, an average of 6–9m apart (thus the gathering is called an 'exploded lek'). Within the gathering of males, only one bird is dominant, and there is evidence to suggest that it tends to have a larger court that the others and undertakes more 'gardening' to remove leaves from his display branches. This dominant bird also arrives first in the early morning and is last to leave at night, and it is this bird that takes responsibility for ejecting any male visitors that might dare to trespass.

Opposite: Male cocks-of-the-rock display communally, a system known as a lek. Females visit in order to pick out an individual of their choice.

The biggest perk for the dominant male, however, is far more important – that is to monopolise attention from any female that happens to come to inspect what the lek has to offer. A female arrival might send every one of the lek males into a frenzy of zealous displaying, but the female ignores all but the top bird, and it is invariably this male with whom the female mates. Evidently any females quickly establish the whereabouts of the dominant male's court, and they unhesitatingly make for that. The lek is designed to sort males by merit and thus guide the visiting female to the best genes.

But in the Andean Cock-of-the-rock, remember, every displaying male has a same-sex partner. These birds actually display in pairs. Males confront each other a few metres apart and bow at each other, jump about and flap their wings, calling all the while. It's the same with the dominant male. The dominant male initiates mock attacks on his partner, and the partner responds. The partner is always the same and, despite the fact that they fight at the lek, the relationship is amicable and the birds are sometimes seen feeding together.

In a fair world you might expect both partners to elicit admiration from the females that watch, and sexual favours to be shared between them. But this doesn't happen. However much the dominant bird's partner displays, it does not mate with any females. Instead the junior partner carries on displaying, or even more remarkably, perches still and watches the dominant male cashing in on the fruits of his labour.

At face value there is no benefit to be gained for the junior partner at all, and so its willing co-operation and passivity is something of a mystery. There is, however, a clue to be had in examining lek behaviour in other species of birds. In some relatives of the cotingas, the manakins, the junior partner is the heir to the dominant male's position. When the dominant male succumbs, the succession goes to this apparently selfless serf. It is likely to be the same in Andean Cocks-of-the-rock. Nevertheless, it's hard to avoid the impression that the partners work hard and for a long time with only hope to play for. It is truly a remarkable relationship.

Opposite: Most unusually, cocks-of-the-rock form coalitions between displaying males. Here two of the birds will be displaying-partners, the other one a rival.

TOUCANS

Why a big bill pays

It's the obvious question to ask. You see a big toucan flying across a river or a clearing on its way to the next fruiting tree. As it flies, its outsize bill, obvious at any distance, seems like a hindrance, and the bird flies with floppy, unsteady and weak wing-beats. When the bird perches, the bill seems awkward and makes its owner look unbalanced. As far as plucking fruit is concerned, it would seem like the proverbial sledgehammer tackling a nut. What is such a bill for?

It almost seems rude to enquire, since toucans obviously invest so much in this part of their anatomy. This feature is the family calling card, instantaneously rendering a toucan unmistakable. Depending on species, it can be a quarter to a third of the length of the bird, 14–15cm in actual measurement. In most species it is brightly coloured, sometimes more so than the plumage, with unusual patterns than have no parallel anywhere in the world of birds. It is big and spectacular, but is the bill a folly? Or does it have multiple uses than aren't clear at first sight?

The first misconception to dash is that the toucan's bill is somehow uncomfortably bulky. In fact it isn't particularly heavy and seems not to unbalance the bird especially; toucans would probably be a little clumsy with or without their bills. The bill itself has a thin, horny outer sheath and most of the inside is hollow; there are, however, many narrow supporting struts criss-crossing this space. But overall, it is light and fragile, easily broken and not quite what it seems.

But what advantage does it confer upon its owner? One definite perk is its reach. Toucans spend most of their time in the tree-tops feeding on fruit and berries of all shapes and sizes. But while most competitors can only harvest fruit close by them, toucans can hold tightly on to their perch and lean down towards the thin outer branches that remain untouched. The bill has forward pointing serrations inside that enable the toucan to get a firm grip on the fruit. The bird can thus tug fruits hard from their attachment and, if necessary, it can also turn its bill one way or another to loosen then. Once it has gathered a fruit a toucan rights itself and, with a flick of the head, tosses the fruit backwards into the interior of the bill. Toucans also have long tongues that gain control of the fruits and transport them to the gullet.

If the bill is such a useful tool, why is it often so bold and colourful? As mentioned above, toucan bills are eye-catching for more reasons than size. The Rainbow-billed Toucan (*Ramphastos sulfuratus*) has bright apple-green

mandibles with a yellow top ridge, together with a reddish-pink tip, a sky-blue sub-terminal spot, an orange streak along the middle, a sharply defined black base and a few dark vertical streaks span the mandibles. Meanwhile, the Lettered Araçari (*Pteroglossus inscriptus*) sports more or less vertical black squiggly lines along the base of its upper mandible, looking like some kind of script. Other araçaris also have curious markings along the lateral line where the mandibles meet, from dotted white lines to black waves following the serrated edge. Some of these distinctly resemble teeth.

In view of the fact that, on average, male toucans have 10 per cent longer and narrower bills than females, you might think that the colour and pattern would vary too, and that the bill pattern would act as a secondary sexual characteristic. However, the bill patterns tend to look very similar between

Above: The bill of each species (and often subspecies) of toucan is unique, often with an elaborate pattern, as in this Rainbow-billed Toucan.

male and female, so unless they show up differently in the ultraviolet spectrum, perhaps this can be ruled out. The bills do, of course, differ between species of toucans, so recognising a rival of the same species is certainly easy. Toucans also have distinctive calls for this purpose, which they use a great deal. At times the forests can ring to their croaks and rattles.

Researchers have come to one firm conclusion regarding a toucan's bill colour and patterning, though – it is intimidating. Doubtless this is sometimes the case when two rival toucans do battle, which they often do, fencing with the bills and grappling. But the real advantage seems to be the reaction of other birds, competitors in the fruiting tree. It seems that everything from pigeons to tanagers simply cowers in the presence of

*Opposite: The sharp hook at the tip of the bill of this Yellow-throated Toucan (Ramphastos ambiguus) hints at the more predatory side of a toucan's diet. **Above:** Male toucans (left) have longer and narrower bills than females, as seen in these Crimson-rumped Toucanets (Aulacorhynchus haematopygus).*

toucans, evidently out of all proportion to the threat that the bill could pose. As a result, toucans live on top of the tree, in every sense. If they wish to drive other birds away from choice fruits, they are able to do so, and thereby gain a competitive advantage.

In fact, although the bill could do little physical harm to most birds, that does not mean than toucans are benign creatures that are entirely bluffing all their lives. As it happens, although fruit constitutes the major part of the diet, a small but significant percentage is meat. People who mist-net small birds in the Neotropics sometimes see a sinister side to toucans, as these bold and colourful birds are drawn in by the distress calls of the captives for the chance of an easy meal. Toucans are opportunists that will eat insects, lizards and small mammals. They also raid the nests of other birds for eggs and young. And for the latter, the bill is a useful tool in two ways.

The first way is obvious. Just as the long reach of the bill is ideal for stretching for fruit, so it is also useful for getting hold of eggs and young inside nests and cavities. The hooked end of the bill is perfect for tearing apart nests, too; various toucans regularly attack the conspicuous hanging nests of oropendolas and caciques.

Not surprisingly, birds don't appreciate their eggs and young being attacked, and on the whole this can provoke a furious reaction among affected parents. Both oropendolas (*Psarocolius* spp.) and caciques (*Cacicus* spp.) are relatively large, substantial and aggressive species. But here, once again, they have no answer to the fearsome sight of the toucan's bill, and keep their distance until the toucan has finished its raid. The bill dampens down the mobbing, and the toucan gets its meal.

Scientists are sure that, at heart, toucans are fruit-eating birds, and that their predatory side has evolved as a sideline, adding some protein to the diet. One thing is for sure. When you watch these bold and colourful birds feeding and ask what the outsize bill is for, the very last answer you expect is that the colour and pattern intimidates parents whose nests the owners are raiding.

*Opposite: The pattern on the bill of a toucan can be intimidating, as well as its size. Note how the bill of this Chestnut-eared Araçari (*Pteroglossus castanotis) seems to have tooth marks.*

ANTBIRDS

Following the ants

An army ant column is a formidable sight, and you need to take a deep breath before you step into it. You wait as the ants in front march in and out of the dappled forest light, and their movement is hypnotic, hundreds of side-streams within the 4–12m wide river of living things, each member of the ant column with an aggressive mission and with plenty of bite. Everything in your guts tells you not to plunge. But you check the powder on your feet and you ensure that everything is tucked in, and then, swallowing hard, you go, swatting the sweat bees away, hoping against insect hell. Sometimes, when we are birding, we have to go extreme.

You need to brave an army ant column if you want to meet a remarkable group of small forest birds which have forsaken everything to be in thrall to these frighteningly predatory insects. These are the professional army-ant followers. They give their name to one of the largest groups of families of birds in Central and South America – the antbirds (Thamnophilidae, Formicariidae), but they are actually not quite all antbirds, and by no means all antbirds follow their lifestyle. They couldn't; it is too alternative.

The ant-followers, such as the Ocellated Antbird (*Phaenostictus mcleannani*) and the White-plumed Antbird (*Pithys albifrons*) really are followers, enslaved by the cult of their own ecology. They spend all their foraging time next to the marauding ant armies, and are simply seen away from them, except at the nest. These insectivorous birds are not so physically adapted to their lifestyle that they could not forage in the 'normal' way, well away from any ants. It's just that they never do. The concept of finding a few stray insects on the otherwise peaceful forest floor is alien to them.

Instead, they follow the violence and menace of army ants, mostly one species called *Eciton burchelli*. It might be an exaggeration to say that army ants on the march spread fear and terror among the ground-dwelling population of invertebrates and small vertebrates, but it certainly looks that way. Just in front of the vanguard of ants, panic ensues. Diverse creatures from tarantulas to beetles and from other ants to small lizards may be seen running or flying for their lives, some taking refuge, some just running. If the ants catch up with an organism they quickly swarm over it, bite it and, if it is large enough, dismember it before carrying it back to the larvae. They are like a plague on the litter fauna, leaving it diminished.

For an ant-follower, the shock-wave of fleeing insects ahead or to the side of the ant columns is the perfect place to forage. Panicked animals of all

kinds make mistakes and are easy to catch. It is the perfect niche lifestyle. Follow the ant-followers and you will see something you are not used to: satiated birds. After a few hours these specialists become full and more and more choosy, and select only their favourite items. The one thing they don't eat is army ants.

It isn't just the odd day here and there that ant-followers forage like this; it is every single day. Scientists call this small guild of birds Obligate Antfollowers because, if you like, this is their profession – they are obliged to do it. They don't wish for, or get a day off. The key to maintaining their lifestyle is that there are always army ant colonies about. In some rich forests there are, on average, three colonies per square kilometre, and antbirds have been

Above: The Ocellated Antbird has a single foraging strategy, from which it never strays. Invariably it catches small animals disturbed by hunting masses of army ants, day in, day out.

●

recorded transferring between several in a day. This is how Obligate Ant-followers can reconcile their foraging style with breeding commitments. They cannot change nest-site, but the constant shifting of the ant colonies ensures that there are always some within commuting distance.

To complicate matters, though, the army ants have a life cycle divided into two phases, one of which offers much better pickings than the other. Every colony alternates between a 'statary' phase when the queen is laying eggs and the workers don't do so much destructive hunting, and a 'nomadic' phase, when the ants launch daily raids to feed the queen's larvae. During the three-week long statary phase the ants bivouac at the same place each night, their own bodies forming the nest, and their hunting raids occur every other day. Once the eggs hatch, however, they enter the 'nomadic' phase, which is exactly that. They set off in any direction on massive hunting raids

Above: Spotted Antbirds are habitual followers of army ant swarms, but they sometimes forage away from the action.

•

with 200,000 or more prowling workers, and set up a bivouac in a different place each night. An antbird will not, therefore, follow the same colonies throughout its life, but many different ones. The 'nomadic' phase of each colony serves the antbird's needs better.

The dynamic and nomadic nature of the ant colony means that ant-following isn't always straightforward. On 'statary' days ants may not hunt at all, or on rainy days; and nomadic ants wander in random, unpredictable fashion. Finding columns that may be 200m long and 20m in width isn't as easy as it sounds on the gloomily-lit litter of the dense forest floor. Obligate Ant-followers begin searching at dawn, hopping and flying a metre or two above the forest floor, calling and singing. They search the sites of overnight bivouacs to assess the ants' activity. Families of ant-followers sing to each other, and even listen for the songs and calls of other species doing the same, until a hunting colony of ants is located. Then 20 or more individual birds may collect at the scene, many staying almost all day.

You won't be surprised to learn that obligate ant-following isn't for everybody; perhaps 20–30 species in all. But equally, the ants provide perfect opportunities for insectivorous birds. So every ant swarm tends to have a coalition of birds present with different degrees of commitment. Some species are very regular followers, but not slavish; at any rate, you routinely see these species, such as Spotted Antbird (*Hylophylax naevioides*), foraging elsewhere well away from active swarms, in contrast to the Obligate Ant-followers, so you could consider them part-timers. Still other species do follow ant swarms, but only occasionally. Highly territorial species might, for example, follow the ants when they enter their patch, but abandon the swarm when it moves on.

Thus, every time you encounter a swarm, you don't quite know what you are going to see. I can recall a swarm on the Peruvian forest floor attended by two obligates, a White-throated Antbird (*Gymnopithys salvini*) and a White-chinned Woodcreeper (*Dendrocincla merula*), plus one regular follower, a Scale-backed Antbird (*Hylophylax poecilonotus*), an occasional attendee in the shape of a Spot-winged Antbird (*Schistocichla leucostigma*), and a Rufous-capped Ant-thrush (*Formicarius colma*) that appeared to be right on the periphery. It was several years ago, and I can still feel the frisson of thrill at the variety of unusual species and lifestyles. It was insect hell, alright – but it was also birding heaven.

●

TANAGERS

The crown jewels

It's just after dawn on a neotropical canopy tower, and a biodiversity bonanza is approaching. It's the hour when the lowland terra firme forest delivers on its promise. The experience of tropical forest birding is often frustrating on the ground; the forest floor is dark, and you crane your neck uncomfortably to see a multitude of flitting shapes high above. But take the ladder up to the tops of the trees in the pre-dawn darkness and the forest cannot hide its crown jewels for long. Soon you will see flocks of multi-coloured birds in their natural habitat, flitting by rapidly, a challenge for the eyes and identification skills. It's the zenith of birding; these are the coral reefs of dry land.

Here in South America, there are plenty of dazzling forms, but you don't need to visit the canopy many times before you realise how many of them are tanagers. This large family of fruit and insect-eating birds exhibits the best catalogue of gaudy wardrobes in the whole of the Neotropics, aside from hummingbirds, and you hear the delight in the describers' pens from long ago as they named them: Beryl-spangled Tanager, Glistening-green Tanager, Seven-coloured Tanager. Unlike hummers, tanagers have a basically similar body form with chunky, almost finch-like build and thick bills, so they impress mainly with their bold patterns and vivid colours. They move quickly in the canopy in mixed flocks, often with other brightly-coloured birds.

If you have a really good tanager day – and 15 species from the family is easily possible – and you reflect on what you have seen, an age-old question might rear its head: How does the forest support so many different bird species? In the case of tanagers, they are clearly all closely related, yet equally they are different. There is an ecological law called Competitive Exclusion. No species can ever share precisely the same niche as another, without one or the other 'winning' and the other disappearing, so long as they are both found in exactly the same part of the world. All these birds, therefore, must have their own niche.

Fortunately for us, a number of scientists have asked themselves the same question, including David and Barbara Snow, Alexander Skutch, Morton and Phyllis Isler and K. Naoki. Their detailed studies have revealed how a wealth of tanager species can coexist (more than 50 species may occur in a single locality) by an extreme and delicate fine-tuning of their ecological needs. There is scope for separation in a number of ways. Tanagers eat both fruits and arthropods (mainly insects), so species of tanagers that eat very

similar fruits might take quite different arthropods, and vice versa. If any species overlap in the fruits they eat, they might locate them in slightly different places, and the same applies to arthropods. If, by any chance, the birds overlap in their entire diet (which in practice is unlikely), they could partition the forest, so that one species feeds in the canopy and the other in the understorey. More or less all these things do actually happen, and that's the beauty of it, biodiversity in action, in real life.

In the wild, most tanagers take a wide variety of different fruits, but there are a number of specialists. One of the most popular food sources among tanagers as a whole is a genus of plants with small fruits known as *Miconia*. *Miconia* is, you might say, the most popular tanager outlet, like a pizza house is for people. However, one particular tanager, the Blue-and-black Tanager (*Tangara vassorii*), appears to feed at *Miconia* shrubs to the exclusion of everything else, while the Emerald Tanager (*T. florida*) takes two-thirds of

Above: Tanagers in Neotropical forests have finely defined foraging niches. The Green-and-gold Tanager tends to forage on broad branches in the high canopy.

its fruit from this type of plant. Other types of plant berries could instead be described as 'niche' foods: they are favoured by a limited clientele. A good example among tanagers is berries from the mistletoe family (Loranthaceae), which appear in the diet of a number of species, but are especially beloved by the Turquoise Tanager (*T. mexicana*).

On the whole, tanagers forage for fruits where they are available, with a relatively limited scope for searching in different locations. Arthropods, however, are much more widely available in different sections of the forest, allowing considerable scope for differentiation. The Isners found that there were several broad categories of foraging micro-habitat, including branch surfaces, the undersurfaces and upper surfaces of leaves, mossy sections, epiphytes, the air, the ground, dead leaves and flowers. Many species would

---

Above: Everybody's idea of a gaudy tropical bird, the Paradise Tanager forages in the high canopy, often in flocks of its own kind.

forage in a number of micro-habitats, but most would have a favourite or specialist location. For example, the Golden-naped Tanager (*T. ruficervix*) catches some 70 per cent of its arthropods by making short flights, sallying to catch its food in mid-air, while the Blue-necked Tanager (*T. cyanicollis*) makes aerial sallies but will also fly towards leaves to catch the insects on or close to their surface. The Yellow-bellied Tanager (*T. xanthogastra*) often hovers. On the other hand, most tanagers don't sally at all, but may primarily examine leaves such as the Spotted Tanager (*T. punctata*) or mossy clumps, such as the Golden-eared Tanager (*T. chrysotis*), or epiphytes, such as the Opal-rumped Tanager (*T. velia*). As far as micro-micro-habitats go, species can be very fussy indeed. The Emerald Tanager usually examines branches between 1.3cm and 2.5cm in diameter; any more or less is out of its niche. The Green-and-gold Tanager (*T. schrankii)* prefers broader branches and the Beryl-spangled Tanager (*T. nigroviridis)* very thin branches. The gorgeous Paradise Tanager prefers bare branches, almost to the exclusion of others.

As a corollary with preferences in micro-habitat, tanagers also differentiate by the way that they obtain their insect food. Sallying is one obvious way, but there are also other, far more subtle differences. The Bay-headed Tanager (*T. gyrola*) practises a method known as a Diagonal Lean, in which it perches on the top of a branch and leans down first on one side, and then the other; it will then repeat this method a metre or so further along the branch. The Blue-and-black Tanager also uses this method, but will also sometimes actually hang upside down.

And, of course, different species forage at different heights. Some tanagers feed on the ground, others in the middle storeys and still others up in the canopy. There is even a sub-group of upper canopy feeders, which includes the Paradise, the Green-and-gold and the Turquoise Tanager. These species feed in the very tallest trees (the emergent trees) of the forest, well above the others. Several of them work the bare branches in small flocks. They are the easiest to see. The sun falls on these colourful birds before all the others.

The biodiversity bonanza in the rainforest canopy is something of a feast of birding, but it is easy to become overwhelmed and to get the equivalent of indigestion. But if you take your birds tanager by tanager, micro-habitat by micro-habitat, you can take your sustenance in more understandable, palatable courses. This is the key to enjoying the wondrous variety of the richest habitat on earth.

HUMMINGBIRDS

When the humming stops

If ever there was a bird for which the abnormal was normal it would be a hummingbird. Superlatives hover around the 340 or so species that decorate the Americas, as much as they themselves hover at flower-heads. Among their number are the world's smallest birds and arguably the lightest vertebrates. They have the highest metabolic rate and oxygen consumption of any bird, the smallest eggs, the fastest wing-beats and the fewest feathers. They are unique in their ability to hover in a sustained fashion and, especially, to fly backwards and upside-down. They attain astonishing speeds. Upon even the briefest meeting you can tell they are something special.

A hummingbird is a small organism in a perpetual state of speed. This is fuelled by a super high-energy fuel, nectar, which the hummers drink neat from the blooms that they visit. The nectar is provided by the plant in return for pollination, the birds invariably getting a dusting of pollen from every corolla they visit, thus ensuring cross-fertilisation. Plants can modify the amount of pollen their blooms provide and exert some control over their pollinators, but must always ensure that the hummers are suitably hooked. Plants that cater to hummingbirds and other birds are known as ornithophilic.

If our children sucked on a sugar drink at the rate that hummingbirds did, we would be horrified. The birds have specially adapted tongues, forked in two, each fork with a narrow trough that allows the nectar to be forced up by capillary action, which is a very rapid process. A feeding hummingbird takes between three and 13 licks per second, ensuring a rate of consumption that would stretch the most liberal of table manners. The liquid hardly remains in the crop or stomach for any serious length of time, but instead is rushed to the small intestine. Remarkably, a hummingbird can start using absorbed sugars within 15 minutes of drinking them in. In the course of a day, your average hummer will drink two to three times its own weight in nectar every day, and this requires a visit to anywhere between 1,000 and 2,000 blooms.

Of course, nobody has ever seen an obese hummingbird, because theirs is such a remarkably energy-demanding lifestyle that almost all the fuel is used

Opposite: A Sparkling Violetear (Colibri coruscans) is frozen in flight, during which its wings may beat 70 times a second. Hummingbirds have fewer feathers than most other birds, which may keep them cool when hovering.

rapidly. Even at rest a hummer's heart beats at up to 600 times a minute (10 times a second), and during physical exertion, such as a territorial chase, this can rise an astonishing 1,000 times a minute. The breathing rate at rest is 300 breaths per minute, which rises to 500 in hovering flight. The hummer's lungs and air sacs are particularly capacious to allow for efficient oxygen exchange.

Despite all the statistics alluding to their metabolic extremes, hummingbirds tend not to give a frenetic, exhausting impression when you watch them in the wild. It is one of their many fascinations that they simply hover from flower to flower, with supreme control and no obvious hurry. Of course, when two hummingbirds squabble they can zoom off at impressive speed (and they are known to reach 96 km/h when chasing) but rarely do they seem to lose balance or poise. One moment they can be perching; the next moment they can simply slip into hovering mode, beating their wings 70 times a second, simply a blur.

There is, however, one side to hummingbirds that few of us ever see or perhaps even consider. These pint-sized endotherms seem so controlled and unstoppable that we cannot imagine that their humming will ever cease. But it does, of course, every single night. The dark curtain comes down on the glitter and brightness, and the hummers endure times of challenge. Night-time is potentially harder to endure for hummingbirds than for other small birds, not least because they have fewer feathers covering their skin than other birds, and they lack the downy feathers that insulate other birds. This is ideal for losing excess heat during the day, but at night it isn't helpful at all.

It is now known that hummingbirds of all kinds – and quite possibly every species – survive the night by becoming torpid. This is very much more than a deep sleep; instead it involves a considerable reduction in the hummingbird's metabolic rate. Until recently torpor was only recorded in a few birds, including the Common Poorwill (*Phalaenoptilus nuttallii*), a nightjar from the southern United States that effectively hibernates, remaining torpid for up

Opposite above: Hummingbirds, such as this Tufted Coquette (Lophornis ornatus) visit between 1,000 and 2,000 flower blooms a day.
Opposite below: The iridescent, glittering plumage of hummingbirds, such as this Black-throated Mango (Anthracothorax nigricollis) is entirely down to the physical structure of the feathers, not to pigment.

to four months at a time. A recent study on 18 hummingbird species of many different genera and all body sizes (2.7g–17.5g) found that they all became torpid at night, regardless of the ambient temperature, which was shifted experimentally between 2°C and 25°C. It seems, then, that it is probably a universal survival strategy of the family.

The changes seen during torpor are manifold. The body temperature drops from 38–40°C to between 18–20°C, the heart rate decreases from 600 to only 50–180 beats per minute and the metabolic rate roughly halves. As a result, the bird saves as much as 60 per cent of the stored energy that it would otherwise use up. There is a considerable downside, though. When torpid a hummingbird is essentially defenceless against predators. It takes at least 20 minutes to restore full wakefulness (I know, I know, this is much faster than a human adolescent), which could never be enough if it was discovered. There is another serious problem, too, which is that a torpid hummingbird is always on the edge of death. The researchers found that, whatever the ambient temperature, the internal fires of the hummer always ensured that it remained above 18°C. If a hummingbird was unable to store enough nectar during the night to keep stoked up, it would die during torpor.

Interestingly, hummingbirds aren't actually torpid throughout the night. They usually are for two to six hours, and then, while it is still dark, they begin to shiver their wings and warm their body up to ambient temperature and above. They then probably get some sleep for a couple of hours until dawn finally breaks.

You might think that the majority of hummingbirds lived in climates where, even at night, it would be reasonably warm, but this is far from the case. Quite apart from the much-loved hummers that spend their summers in the temperate regions of North America, a considerable percentage of the species in the family live at high altitude – up to 50 per cent of all species. Indeed, it could be argued that the hummers of the mountains include many of the most outlandish colours and body-forms in the whole family. Some of these live permanently in conditions that would be challenging enough for large mammals such as people, at altitudes above 4,000m. Many hummingbirds see daily ranges between night and day of well over 15°C.

It is a truly humbling experience to visit the grasslands of the páramo in the high Andes, where it is cold, wet and often foggy, to feel dizzy

with altitude and overwhelmed with cold, and then to come across a hummingbird. I have seen a bedraggled looking Blue-mantled Thornbill (*Chalcostigma stanleyi*) at about 4,000m in the early morning, and the Andean Hillstar (*Oreotrochillus estella*) can occur at 5,000m.

But let's face it, hummingbirds are astonishing. Nothing is particularly normal about them, not even at night.

Above: The side of Hummingbirds that we don't see:
an Andean Hillstar pauses for a rest. Hummingbirds go
into torpor at night, and sometimes during the day, too.

•

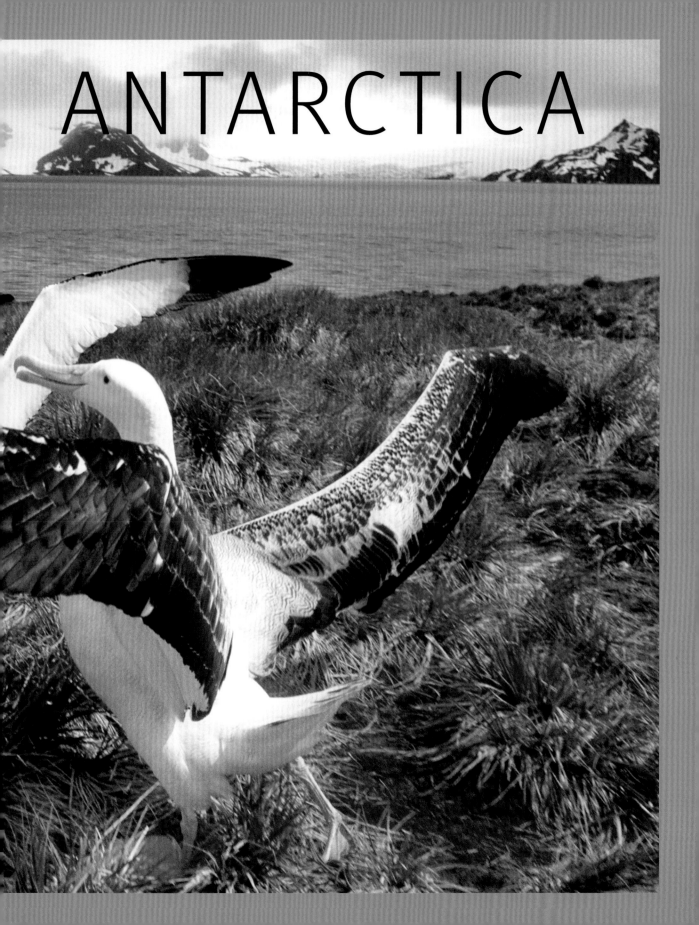

ANTARCTICA

ROCKHOPPER PENGUIN

The most unloved egg

Visiting a penguin colony in the Falkland Islands (*Islas Malvinas*) off the southern tip of South America is like walking through your own documentary programme. It is almost impossible not to imagine the commentary as you stop to observe snapshots of life: a pumped up Gentoo Penguin (*Pygoscelis papua*) holding its head to the heavens and trumpeting loudly; long lines of Magellanic Penguins (*Spheniscus magellanicus*) marching en masse towards the ocean, as if taking part in a very slow city marathon race; King Penguins (*Aptenodytes patagonicus*) standing awkwardly but determinedly with their precious egg upon their feet. You will have heard the sound-bites countless times: penguins problem in our culture is over-exposure.

In those same documentaries there is another sound-bite that you are also highly likely to have heard whenever some unfortunate seabird has succumbed to a fatal mishap. Alongside footage of a Falkland Skua (*Stercorarius antarcticus*) or a Snowy Sheathbill (*Chionis albus*) consuming the bloody body of a departed seabird, the mantra is always the same. 'In this harsh environment,' it intones, 'nothing goes to waste.' And down the throat of the scavenger goes this sorry piece of organic collateral.

And yet, a closer look at the life of one Falkland bird suggests that those casual fragments of commentary are, at least in one case, a long way off the mark. It would appear that the fourth species of Falkland penguin, contrary to sensible biological principles, is in fact very wasteful indeed. Every season, without fail, every fertile pair of Southern Rockhopper Penguins (*Eudyptes chrysocome*) lays two eggs. But in every case, the sum of their production is invariably one chick. In the vast majority of cases, only one egg actually hatches at all, leaving the other fallow. The question has baffled scientists: why lay two eggs, when there is no chance that you will ever raise two young?

It should be pointed out that it isn't especially unusual in the world of birds for an egg to be the fall-guy in a clutch. There is a brutal but effective strategy operated by a good number of species of birds, especially those that depend on unpredictable resources, and it's called Brood Reduction. The idea goes like this. A bird lays two or more eggs over the course of a few days but, in contrast to the majority of bird species, incubation begins with the first egg. The first-laid inevitably becomes the first-born, the oldest and, as long as it is healthy, the largest, conferring upon it a competitive advantage over its siblings when begging. If resources are plentiful, it becomes intermittently satiated and its siblings get a chance to feed, and survive, producing a great

breeding result. If resources are scarce, the first-hatched still gets first pick of any food and consequently, may well still survive at the expense of its siblings, producing a decent breeding result. In the latter situation, if the parents were to distribute food evenly, resources would be spread thinly with the result that all in the brood might starve, which is a disastrous breeding result. Uneven provisioning is often seen as an insurance policy to make sure that at least one chick is produced.

Above: A Rockhopper Penguin incubating two eggs.
Almost invariably, only one of them ever hatches.

The very fact that Brood Reduction makes sense magnifies the fact that this penguin's version doesn't make sense. In this species, remember, Brood Reduction cannot strictly happen because there isn't a brood to reduce – usually just the one egg hatches. Even if both eggs hatch, one chick always dies within a few hours or days of brooding. The reason for this is extraordinary: the successful egg is invariably much larger than the other, to the tune of 85 per cent, almost double the weight. You might be persuaded that the inadequate embryo of the smaller egg barely has a chance. As such, you cannot contend that laying two eggs instead of one is in any way an insurance policy, because the two eggs are so unequal.

Another aspect of the rockhopper's laying is still more unexpected. Astonishingly – and entirely counterintuitive to Brood Reduction theory – it is the first-laid egg, not the second, that is smaller and doomed to failure. And this is still true despite the fact that it is laid, on average, 4.4 days before the clutch is completed. With such a huge head start, one of the biggest between siblings among any birds in the world, you might think that the chick from the first egg has a significant advantage. However, this potential inherent opportunity is squandered.

We return, then, to the original question. Why do Southern Rockhopper Penguins lay two eggs when physiologically they cannot support bringing up two offspring? Other species of penguins, such as the Magellanic and Gentoo Penguins that share their home in these South Atlantic islands, do manage to bring up two young. So why are the rockhoppers unable to do so? The truth is that, despite much speculation, nobody has yet come up with a fully satisfactory answer.

There are, though, two interesting strands of thought on this. One suggests that these rockhoppers are in the process of evolving into penguins that invest all their energies into a single egg. Today there are two species

---

Opposite above: A Rockhopper turns an egg. The second-laid egg is always much larger than the first one, and it is this one that the adults actually incubate properly. Opposite below: A group of juvenile Rockhopper takes a rest. Only occasionally do Rockhoppers ever raise siblings.

of penguins in the world that do this, the King Penguin and the Emperor Penguin (*Aptenodytes forsteri*). Both are, like the rockhoppers, offshore foragers, travelling a long way to collect food for their single youngster. It has been speculated that, at some point in their evolutionary history, these larger species dropped the second egg, so to speak, departing from the ancestral two-egg clutch. Perhaps the rockhoppers are on the same path?

Perhaps, and studies into the closely related Yellow-eyed Penguin (*Megadyptes antipodes*) of New Zealand add an intriguing dimension. In this species, the laying of the first egg acts as the stimulus for the development of the brood patch used for incubation. This is a way of getting around having a large age gap between chicks, and consequent uneven competition (as in Brood Reduction). In this species, which lays eggs of roughly the same size, proper incubation begins only when the brood patch is complete after the laying of the second egg. The result is that both hatch at the same time.

It could well be that rockhoppers once did this, but over time, perhaps as they become adapted to feeding further offshore, they could no longer afford to raise two chicks. Over time the first egg retained its function of stimulating brood patch production, but lost its viability in favour of the second egg, which always came along when the brood patch was 'ready'. This is where we are today, on the path towards a clutch of one.

And that, on the face of it, is quite a waste of an egg.

Opposite: An adult broods a well-grown chick.

●

ALBATROSSES

Masters of the oceans

It is difficult to see an albatross without emotion. That applies even where there are plenty. I myself have marvelled at them at the breeding areas in the Falkland Islands, and seen hundreds passing the western coast of Kangaroo Island, off South Australia. Every one of these large, streamlined and long-winged birds was a jolt of birding electricity. The size of albatrosses makes them command your attention, especially if they are accompanied by shearwaters at the same time, birds that are hardly small but are nonetheless dwarfed by the big gliders. The shape of albatrosses, together with their effortless wheeling flight, with barely a wing-beat, is what thrills your heart. There are no birds like them.

Everybody knows that albatrosses are good at flying. That very shape, with long, narrow, pointed wings is adapted for energy-saving progress in the blustery winds of the Southern Ocean, where albatrosses are most abundant. Their wings are longer than most other birds, up to 3.5m in the case of the Wandering Albatross (*Diomedea exulans*). They also have more flight feathers than the rest, including 25–34 of the secondary feathers that form most of the trailing edge of the wing (a small bird has 10–20). The great birds have special tendons that fix the wings into the fully extended position. All this ensures that each wing is an efficient aerofoil, ideal for making fast forward progress without slowly sinking. Albatrosses on the oceans utilise two different types of energy-efficient flight: dynamic soaring and slope soaring. In dynamic soaring they dive into the troughs between waves and then pull upwards into the wind; as they do they hit a headwind and flip over, using the wind behind them to dive fast back into the lee of a wave, and so on. They therefore use the energy gradient of the wind to power them. In slope-soaring, the birds approach a wave and take advantage of the updraft where the wind hits this water obstacle. Both types of soaring allow the birds to travel while expending very small amounts of energy.

The flying skills of albatrosses have been known for many years, and universally admired. Scientists and authors have long speculated on just how good they actually are. The question as to how far albatrosses might wander on the ocean was often asked, but only guessed at.

But now we know, and the answer only adds to the mystique of these birds. In the last 20 years, we have tracked them with data-loggers in contact with satellites. The results have amazed everybody. It seems that, for albatrosses, the world – or at least, the Southern Hemisphere, is one large playground.

Take the albatrosses passing Kangaroo Island mentioned above, which were Shy Albatrosses (*Thalassarche cauta*). If these were typical of the species on migration, satellite tracking studies suggest that, by the end of an average flying day, a Shy Albatross could well be 900km away from where it started. Six individual Wandering Albatrosses revealed the sort of distance that can be travelled by post-breeding birds. One individual journeyed 10,427km over 27 days, and had a top recorded speed on 63 kilometres per hour. Another individual moved 15,200km in 33 days, including 936km in a single day. This bird managed a top recorded speed of 81 kilometres per hour, and one gets the impression that higher speeds were possible.

Above: An adult Grey-headed Albatross guards its chick. The other parent will be out at sea looking for food, perhaps thousands of kilometres away.

Grey-headed Albatrosses similarly tracked have been no less impressive.
A number of individuals breeding on South Georgia, in the south Atlantic,
were satellite tracked after breeding (most albatrosses take a sabbatical, only
nesting in alternate years). One bird travelled from South Georgia to a patch
of water in the south-west Indian Ocean, taking only 6.2 days to get there at
an average of 950km every day. Other individuals did what everybody does
after breeding (don't they?) and breezily circumnavigated the world. One bird
accomplished this feat in only 46 days. Males may well circumnavigate the
world several times during their year off – an incredible thought.

*Opposite: The long, narrow wings of an albatross such as this Light-
mantled Sooty Albatross allow it to travel enormous distances while
expending negligible energy. When migrating, an albatross may cover
over 900km a day. **Above:** Much as albatrosses are renowned for their
flying prowess, they are very much seabirds, and often swim.*

Most birders are aware that the longest regular migration is carried out by the Arctic Tern (*Sterna paradisaea*), which can travel from the Arctic Circle to the Antarctic (shortest distance between the two 19,000km) in a journey that may take it up to 90,000km in a year, including considerable meanderings in the Antarctic. If you compare this to an albatross's wanderings, it doesn't become trivial, but it is nonetheless put in the shade. A study on 13 tracked Wandering Albatrosses found that they moved an average of 184,000km in their first year of life. Only a few years ago such distances were considered impossible.

Albatrosses even cover enormous distances when feeding their young. Both sexes have this task, and they tend to alternate 'short' trips of one to three days away with much longer ones of five days or more. One male Wandering Albatross on South Georgia made two trips totalling 9,280km – and all to find food for the chick waiting on the nest.

Finding food, incidentally, on the water surface is not necessarily easy, and part of the reason for the ocean-wide peregrinations of albatrosses is that they are seeking resources that are distinctly patchy. To that end, they seek out sites where currents rise, where waters meet and even where whales are feeding. They often follow trawlers, too, sometimes at the risk of becoming drowned on long lines, being caught on hooks meant for fish. Non-breeding birds often return to the same patch of ocean each year after breeding.

There is one, quite unexpected aspect of albatross natural history that has also been uncovered recently, and merely adds to the list of talents possessed by these birds. Albatrosses principally eat fish and squid, which they tend to get by surface-seizing, reaching down from a swimming position. In the case of one of the great albatrosses, such as Northern or Southern Royal Albatrosses (*Diomedea epomophora*) and (*D. sanfordi*), the bills have a reach of about one metre, so they take surface-swimming fish and squid, along with some krill and carrion. However, perhaps surprisingly, it now turns out that some albatrosses can go fairly deep underwater. In a study of Shy Albatrosses, it was shown that they do two types of dive. One resembles the plunge-diving of gannets and boobies (Sulidae) from a height, and can reach down to about three metres. The other type is the swimming dive, and this recently discovered talent can take one of these large birds down to 7.4m, the deepest recorded for this species.

But the Shy Albatross is not the only swimming diver, able to pursue prey well underwater. In recent years the Light-mantled Sooty Albatross (*Phoebetria palpebrata*) has been recorded at depths that averaged 4.7m and stretched to an incredible 12m, presumably chasing squid.

It turns out that albatrosses are masters of the ocean alright – but both above and below the surface.

Above: An albatross usually snatches prey while swimming on the surface, but some species can dive as deep as 12m below the surface.

EMPEROR AND GALAPAGOS PENGUINS
A tale of two penguins

The Emperor Penguin is a species of extremes; you don't need to be a passionate birdwatcher to know that; almost everybody does. It is a multiple record-breaker and professional superlative and features in so many wildlife books that it is a celebrity in the animal world. It's the deepest diver of all birds, plunging down to 535m on occasion, into a realm very different to anything we experience. It is the only bird species to breed during the Antarctic winter, when it has to contend with ferocious polar storms. The males go the longest without food of any species (115 days) except perhaps, the Poorwill of the American desert, which hibernates. It is one of only three bird species to feed its young on a type of milk, made in the crop. The Emperor Penguin is so extraordinary that its freakiness becomes a cliché. Sometimes an animal goes so far out on a limb that any yardstick we can measure it by is hopelessly left behind. When pre-eminence is in a vacuum, it loses its meaning.

The Emperor Penguin, handsomely turned out in its designer black, white, grey and apricot and set perfectly against the shades of polar sky and ice, has come to typify penguins as a whole in the public imagination. And that is what makes it feel so very odd – discordant even – when you encounter the Galapagos Penguin (*Spheniscus mendiculus*). To say that these two species are poles apart is hardly correct geographically, because the Emperor Penguin never reaches the South Pole and the Galapagos Penguin actually lives on the Equator, where a small number of individuals nudge the species, and the penguin family, into the Northern Hemisphere. But the sheer difference in the two species, in their habitat and life history, is so stark that they stretch the boundaries of family connection further than any other species pair in the world. And helpfully, it puts the escapades of the Emperor into some sort of context.

Even the briefest encounter with a wild Galapagos Penguin jolts you away from the typical image of the family. The chances are that you will see it on an islet, or rocks by the ocean, where it will be standing on hard, black lava flows. This, of course, is in contrast to the white world of the Emperor Penguin, the ice and snow it sees all year round. That is the difference between the rock of a Galapagos Penguin and the hard place of an Emperor.

However, the most unexpected feature of the Galapagos Penguin experience is the sun on your back. It isn't just the normal sun, but the equatorial, broiling sun, stifling and exhausting. Under these conditions

the Galapagos Penguin lives and breeds, and the stress often shows. Loafing Galapagos Penguins have trouble keeping cool, and they spend much time panting; to steer clear of the sun, they secrete their eggs and chicks in a burrow or crevice. The temperature on the archipelago tends not to rise much above 30°C, owing to the cooling effect of the nearby Magellanic Current, but incubating Galapagos Penguins have been recorded struggling in 40°C heat, when it would be warm enough literally to fry an egg on the rocks. A word of understanding is required here. These equatorial penguins

Above: A crèche of well-grown Emperor Penguin chicks.
They are brought up in an environment of ice and snow.

aren't masochistic, but live on the islands to take advantage of the cold and nutrient-rich Cromwell Current which rises from the oceanic depths and flows past the western part of the island group, providing a rich source of food all year round. Once in the water, Galapagos Penguins look at home; it's only on the islands themselves where they look as out of place as hapless captive penguins in a zoo.

Galapagos Penguins have two breeding seasons, one of which is June– September. This happens to coincide with the breeding season of the Emperor

Above: Emperors have to endure arguably the harshest conditions of any birds in the world. Temperatures may drop to −60°C in the midst of the Antarctic winter.

Penguin down to the south. The single egg of the Emperor comes along in May or June. The female lays it on to her feet, and there is an awkward moment when she must transfer it to the feet of her mate. This piece of penguin football must not be taken lightly, because if the egg tips on to the ice, it will quickly freeze and be lost; with the male, in particular, having acquired a thick layer of fat to protect him from the cold, is effectively wearing a heavy coat, making foot dexterity especially tricky.

Once the egg is upon his feet and covered with his belly fat, the male's task is simply to endure the extreme conditions. Much has been written about the Emperor Penguin huddle, in which the males group together to keep warm, and have a rota of duty on the outside rim of the group. Not only is it cold and dark, but the violent Antarctic gales sweep through regularly and sometimes plunge the temperature down to -60°C. This means that two related penguin species could be experiencing, at the very same moment, incubating temperatures that are 100 degrees apart.

Meanwhile, on the Equator, the Galapagos Penguin sometimes doesn't need to incubate at all, such is the heat around it. Another difference is that both sexes take turns to incubate. And there is no need for games with the feet. It would be difficult, because in complete contrast to the Emperor Penguin, the Galapagos Penguin usually lays two eggs. This is possible because the two parents take shifts foraging out in the ocean for food. They often don't need to go far to find small fish, and they often return within a couple of days. The eggs are laid a few days apart and the older chick gets the first of the spoils, but in years with plenty of fish the penguins can raise both chicks.

In contrast to the Galapagos Penguin, the delivery of food to an Emperor Penguin chick is much less frequent. After an incubation period of 62 days (38–42 in Galapagos Penguins), the chick is usually fed for the first time when the female visits to relieve the male, although the male may first feed it on a curd-like 'milk' produced in the oesophagus. It is next fed about 25 days later, and then subsequently at shorter intervals. In contrast to the inshore-feeding Galapagos Penguin, Emperors regularly swim up to 500km from the colony in order to find enough food for their youngster. They may also have a walk of up to 120km from the edge of the pack-ice.

The Galapagos Penguin chicks fledge when they are about 60 days old. By this time the Emperor Penguin chicks have gathered into creches and are no

longer brooded or guarded by the adults. On the infrequent occasions when the adults visit, parent and chick recognise each other in the melee by voice.

In common with their male parent, Emperor chicks in a crèche quickly learn to herd together to protect themselves from the lashing storms of the Antarctic. They could be doing this, subjected to some of the worst of the world's weather, at the very same time that the Galapagos Penguin youngsters are finally independent and swimming in the cool waters of the Cromwell Current.

If ever a snapshot illustrated the dichotomy in the tale of two penguin species, this would be it.

*Opposite: The Galapagos Penguin lives on hot volcanic islands on the Equator, even creeping into the Northern Hemisphere. In the wild, these penguins never see ice or snow. **Above:** In complete contrast to the Emperor's frozen habitat, the main danger for Galapagos Penguin eggs (here being incubated) is overheating.*

SHEATHBILLS

The basement cleaners

Do you know what the only bird family confined in breeding range to the Antarctic region is? It isn't penguins, which have temperate and even tropical outposts. Neither is it albatrosses, or any other bird of their ilk. The answer happens to be the sheathbills (Chionidae), a peculiar and obscure family of white birds that look like a cross between a gull and a crow, but have elements of chicken thrown in. When they fly, they look like pigeons.

Sheathbills are landbirds living in a world dominated by seabirds. One species, the Black-faced Sheathbill (*Chionis minor*) lives on the sub-Antarctic islands, Crozet, Kerguelen, Marion and Prince Edward Islands, Heard and McDonald Island. The other species, the Snowy Sheathbill (*Chionis albus*), occurs on the Antarctic peninsula and migrates to the Falklands and southern South America. They are the only birds breeding in the Antarctic that lack webbed feet. They are relatively small, dumpy and unspectacular. In habitats dominated by the world's greatest fliers (albatrosses) and the world's greatest swimmers (penguins), sheathbills aren't the world's greatest at anything.

However, there is one job that, on the whole, does not require strict qualifications, and that is the garbage collector, or dustman. Sheathbills belong to a category of birds known as cleaner birds, the ones that do the waste collection by putting it into their stomach. Vultures do similar things in many parts of the world, although at least they are supreme at soaring flight. Sheathbills go about their business at the feet of their 'clients'.

Actually, there is a qualification needed for every garbage collector, and that is a resistance to being squeamish. The list of items that sheathbills eat is, to put it delicately, broad. These birds are basically professional scavengers, taking dead meat of all kinds, often rather small scraps left over by the larger scavengers of the Antarctic, such as skuas (*Stercorarius* spp.), Kelp Gulls (*Larus dominicanus*) and giant petrels (*Macronectes* spp). This means that they will also eat the placentas of seals and the feather shafts of moulting birds. Away from the meat option, sheathbills are well known for their habit of eating faeces, much to the disdain of Antarctic researchers. They consume without hesitation the nasal mucus of seals and occasionally steal milk from a suckling mother. If a seal or penguin is injured they will feed off the blood, sometimes causing much distress to the unfortunate animal. In the Antarctic, nothing should go to waste.

At this point you might find the sheathbill lifestyle repellent, but at least you can say, as so often is stated in support of vultures, that they play a useful role in the ecosystem, removing waste. That is true with sheathbills, but only partly. Unfortunately, their adaptability goes beyond the role of the rubbish consumer, and crosses the line into unrestrained parasitism and predation. Sheathbills can make penguins' lives a real misery.

The arrival of penguins at their breeding colonies in the spring triggers a time of plenty for sheathbills. Indeed, in many parts of their range they could not breed at all were it not for the unintentional largesse of the flightless birds. Pairs of sheathbills resolutely defend a territory that encompasses a

Above: Sheathbills are almost entirely dependent on penguin colonies for their breeding success.

number of penguin nests, and they pass the time essentially abusing their unfortunate neighbours. They are experts in stealing eggs from under the penguins' noses, especially the semi-neglected eggs of Rockhopper Penguins mentioned earlier (page 166). In more sinister fashion they will also take very young penguin chicks, which they steal away back to their own nests to dismember and eat. This, though, is an occupational hazard for all penguins, which suffer just as much at the hands of skuas and giant petrels.

Indeed, the reactions of penguins to their larger persecutors is invariably far more violent than it is towards sheathbills. Skuas and giant petrels are

Above: The sheathbills are the only landbirds in Antarctica, and are also the only family of birds confined in their breeding range to the Antarctic.

●

physically dangerous to penguins, whereas sheathbills, owing to their much smaller size, are not. This is why the smaller birds often get away with their 'petty' predations and can walk around a penguin colony without inducing a mob reaction. They are also very agile and can escape from a physical assault, either from a penguin or, less often, when they steal the eggs of cormorants and petrels opportunistically.

While sheathbills are not dangerous to adult penguins, they can be, as mentioned above, a severe annoyance. Sheathbills don't just make their nests among penguin colonies for the occasional meal of an egg or chick, or even to clean up penguin mess. The real reason why sheathbills attend the flightless birds is to intercept deliveries of food to the young. When a penguin returns from a foraging trip, it will often be carrying a bolus of highly nutritious food, such as krill, a super-abundant Antarctic crustacean. For a sheathbill, this is a big opportunity. If it can harass a parent penguin into regurgitating the meal, or distract the receiving chick, the sheathbill can thieve the offering for its own chicks. Sometimes a pair of sheathbills will act in concert, one bird annoying the adult by flying up towards its bill and being in close attendance, the other bird keeping the chick occupied. This struggle can be quite physical, and sheathbills have been known to knock penguins over in their desperation to steal food.

On the whole, sheathbill attacks do not actually hurt penguin offspring, because only about one per cent of food brought in to the colonies ends up in the stomachs of the parasites. On the other hand food-stealing, or kleptoparasitism, is crucial to the sheathbills; about 90 per cent of a chick's food may come from this source.

The departure of penguins and seals from their breeding sites at the end of the season marks a downturn in fortunes for sheathbills. Gone is the easy, abundant food; no dead bodies, no eggs, no blood; not even dripping noses. In the winter sheathbills have to knuckle down, eating scraps on the seashore including algae, limpets and other molluscs, and insect larvae. A few individuals even take to feeding inland, out of sight of the sea. On moorland and bogs they subsist mainly on worms, along with other soil invertebrates, often feeding in driving rain or snow.

That, though, is the lot of the lowly. But on the other hand, you could also say that, as far as adaptability is concerned, sheathbills have few peers anywhere. When it comes to making do, sheathbills are kings.

•

WANDERING ALBATROSS

A slow dance to success

The Wandering Albatross (*Diomedea exulans*) is one of the world's supreme fliers. Yet on land it looks ill-at-ease. It walks with its head down and back hunched, and has a rolling gait, like an overfed human builder pushing a wheelbarrow. When it comes to selecting a mate, you might think that the Wandering Albatross could play to its strengths. Perhaps it could impress with a high-speed flight-display, showing off its strength and power and wheeling past the bill-tip of a spectator? Perhaps it could show off its endurance, keeping its flight going in endless circles around the breeding islands? Perhaps it could turn upside-down or do other aerobatics? All of these would, to us, seem a perfect fit.

What does the Wandering Albatross do instead? It dances. It displays on the ground, where its awkwardness is evident, at least to human spectators. Albatross partners choose each other for what, in the broadest sense, they are not good at.

It is hardly a dance. The display is a series of gestures, accompanied by loud braying calls, usually performed with the bird standing still. The reason that it is often described as a dance is that, in common with many a human dance, it is typically a communal activity, with sometimes 20 or more birds taking part at the same time. And in another human parallel, it usually takes two or three brave souls to begin before the rest follow suit.

To see these monstrous birds (and they really are intimidatingly big) indulging in ritualised postures, which they will often continue even when people are in close attendance, is one of those experiences which is spellbinding and a little hilarious at the same time. Scientists and authors have often crept away from a strictly neutral approach when naming the gestures, calling them things like 'Gawky Look' and 'Scapular Action'.

The braying and squealing just add to the fun. One moment the birds are holding their heads up and pointing to the sky, the next moment they are bowing down in courtly fashion. At the same time one bird might be snapping its bill many times in succession ('Yammering'), often aggressively at another bird, while another is performing a subtle ritualised preen.

Opposite: Albatrosses form long-term pair-bonds, lasting a lifetime.

The pinnacle of Wandering Albatross display is Sky-Calling, in which the bird points to the sky and opens its wings, holding them up in a heraldic posture. At the same time it spreads and lifts its tail, and lets out a brief scream lasting two or three seconds. This often provokes a response from the audience of other birds, and may set off a scene of relative chaos. Such displays usually take place at a nest site, on a half-built crater that the bird can use as a stage. Most Sky-Calling is performed by males and, if a female is suitably impressed, she might allow the male to lead her away from the mob of other birds – although this is as much a part of the ritual as the rest and only rarely ends in copulation.

These displays have delighted adventurers and scientists who have come to the Wanderers' remote breeding islands, which are mainly in sub-Antarctic waters. To the albatrosses themselves, of course, the displays have a serious purpose, that of selecting a mate from the various options present. What is perhaps most surprising, though, is just how long the birds take to make their choice. A bird may visit the islands at the very beginning of the breeding season (usually in November) and have a whole summer of displaying without anything getting serious. In many cases it may take six or seven years before a bird finally actually pairs up with anybody and starts a breeding attempt. Interestingly, once they do pair up, Wandering Albatrosses no longer take part in any exaggerated displays at all. Perhaps we shouldn't be surprised; human parents don't, on the whole, go clubbing.

One thing that is not always appreciated about albatrosses in general is just how slow their life cycles actually are. They are extremely long-lived birds if they make it to the end of their first year, and it is likely that the larger species commonly live to 50 or beyond. In these circumstances there is no rush to do anything. And the Wandering Albatross, one of the world's fastest and most awe-inspiring fliers, does little else at any speed.

On average, a Wandering Albatross first visits the breeding grounds after a long period of youth, often five to seven years. When they first arrive the

Opposite: The famous sky-calling display of the Wandering Albatross is only performed by unpaired birds.

young albatrosses are shy and clueless, but quickly settle into the dancing 'game', increasing the number of informal 'pairings' (leading a female away/being led) that they make year on year. Both sexes do not normally breed until they are 10 years old, and occasional individuals are still unpaired at 15. This exceeds the average lifespan of many a smaller species of bird.

Why do they take so long? It can only be assumed that they have to be extremely careful. Albatrosses in general are faithful to their partners, at least partly owing to the fact that the both parents are needed to feed the chick. They are also very slow reproductively. It takes a full year from the laying of an egg to the departure of a fledged chick. Each individual albatross usually only breeds once every other year. You might say that, when they are dancing, albatrosses play for high stakes.

In contrast to most birds, where an individual might engage in copulation outside the pair bond, albatrosses don't do this, and their partnerships tend to be exclusive and life-long. Divorces do occur, but they are very rare and damaging. They only occur after a succession of several failed breeding attempts, and when they do occur, they wipe out an individual's chances until at least the following year. It has been calculated that a divorce can reduce an albatross's lifetime reproductive output by up to 20 per cent.

The albatross dance, therefore, is a delicate, careful and serious business. What the albatrosses are looking for in each other we shall never know, but it doesn't seem to be speed.

Opposite: Well established pairs don't need to enter into elaborate displays – just some calling will do.

●

ISLANDS

SWALLOW-TAILED GULL

Making the most of dark nights

The Swallow-tailed Gull (*Creagrus furcatus*) spends most of its life in tropical waters. Its main breeding station is the Galapagos Islands, which straddle the Equator to the west of continental South America, and there is also an outlying colony on an island off Colombia. This is unusual, because very few gulls live in the tropics, something many birders notice when they arrive for the first time. There are terns, boobies and frigatebirds, but the seabird population is distinctly lacking in gulls.

One of the reasons for the dearth is that tropical waters are far less productive than waters at higher latitudes, so it is harder to find food, especially if, like a gull, you are unable to dive deep. This is particularly the case for the pelagic, very deep waters far from land where the Swallow-tailed Gull mainly feeds. Food distribution is less predictable, ocean food supplies less structured. That is why the greatest density of gulls is in temperate waters at higher latitudes.

There is, however, one aspect of the oceanic cycle on tropical waters that is more reliable than most, and that is known as the Diel (or Diurnal) Vertical Migration, or DVM. This describes the regular movement of organisms from deeper depths by day towards the surface at night. DVM has been described as the largest migration by biomass in the world, and it takes place in all oceans and many large freshwater lakes. Representatives of every type of zooplankton are known to take part and the 'migration' ranges from a few metres to hundreds of metres. On the whole, the animals head to shallower waters at dusk and plunge deeper at dawn, meaning that, in extreme cases, the water surface may have 1,000 times the density of food at night than during the day. The reason why the organisms migrate is probably a simple one; to avoid predation by diurnal foragers.

For any predator on the oceans, it would make sense if they could somehow cash in on DVL to their advantage. And it seems that the Swallow-tailed Gull manages to do just this. At any rate, it has evolved a lifestyle quite unlike any other member of its family – it has become nocturnal, coinciding with the enhanced abundance of food at the surface. At the colonies on cliffs on the Galapagos, the birds become excited and animated at dusk, calling and milling about before they head out to sea in flocks as darkness falls.

Opposite: The Swallow-tailed Gull is unique among gulls for foraging almost entirely at night – but not when the moon is bright.

The Swallow-tailed Gull shows several adaptations to a nocturnal lifestyle. The eye is slightly larger, at least in terms of its optical axis and the cornea, than it is for gulls of comparable size; it also possesses a tapetum, a mirror-like layer of tissue behind the retina which reflects light back outwards, allowing an enhanced stimulus to the photoreceptors. Interestingly, the blood plasma of Swallow-tailed Gulls also shows a more or less constant melatonin level in each 24-hour period, suggesting that the birds don't exhibit a regular sleep pattern. Enhanced melatonin induces a period of rest. Swallow-tailed Gulls may hunt at night, but they are also regularly awake by day. They may fly over the surface of the ocean in daylight, preen on water or on land, and take part in breeding activities, including mating, by day. So, while they seem to be primarily nocturnal foragers, they are not strictly nocturnal animals.

A study carried out recently has added a most intriguing twist to the Swallow-tailed Gull's nocturnal foraging plot. Conducted on 37 breeding pairs in Galapagos, the study, which used Global Location Sensors fitted with a wet/dry component so that the scientists would know when the birds were swimming on the sea, showed that the gulls did not forage every night, but only on some nights. They were, if you like, nocturnal part-timers. There were occasions when the Global Location Sensors detected them at the breeding cliffs after dark, and other times when they would just be flying about at night, and not settled on the sea. This is in some ways a surprising result, because when birds were breeding, and would presumably have chicks to feed, you would expect them to take every opportunity to go out foraging. But on some nights they are feverish foragers, while on others they simply sit still.

It turned out, however, that the Swallow-tailed Gulls were being decidedly pragmatic. They tended to stay put when strong moonlight lit up the surface of the sea. Evidently, the costs of nocturnal foraging in moonlight were greater than the potential reward. The number of birds foraging, and the amount of time they spent on the sea during the night, followed closely to the phases of the moon, with increased foraging at New Moon as well as a general lethargy at and around the time of Full Moon.

It has long been known that daylight is not the only determining factor for DVM, but that it also follows the lunar cycle. Just as planktonic organisms eschew the daylight and migrate to the depths for safety during the day, they stay put in deeper waters on strongly moonlit nights, too. The traffic involved

in the usual surface-oriented DVM greatly reduces, although it does not stop completely.

The favourite foods of Swallow-tailed Gulls are the Purple-back Flying Squid (*Sthenoteuthis oualaniensis*), which is about 10cm long, and small fish in the family Clupeidae. The bird catches them by a technique known as surface-plunging, which involves a forward lunge into the water from a swimming position. The hunter can therefore only catch food when it comes right towards the surface of the ocean, and both the squid and shoals of clupeid fish are known to show a strong tendency towards DVM. With its limited fishing technique, you might say that the Swallow-tailed Gull is at the sharp end of DVM variation. It would seem that, unless there is significant nocturnal DVM, it simply doesn't pay to forage at all.

The Swallow-tailed Gull is the only gull in the world that principally forages nocturnally, although it is by means the only seabird to do so. The research done on its adaptation to the lunar cycle shows an impressive flexibility, in addition to its switch to foraging in reduced light. In future it is likely that other pelagic seabirds will be found to show a similar aptitude.

Above: Large eyes are a clue to the Swallow-tailed Gull's nocturnal habits.

MEGAPODES

The patter of great, big feet

Bigfoot is alive and well and living on islands in the Pacific Ocean, but this one is a peaceful creature that uses its feet for digging rather than engaging in any violent activities. The news will disappoint those advocating ape-like creatures in North America, but in terms of well-proven biology, the real Bigfeet have their own intriguing story to tell, delving back into a time when plenty of megafauna stalked the earth.

The modern-day Bigfeet are a small family of chicken-like birds that are usually known as megapodes. The name arises from the generic name *Megapodius* and this, of course, translates as 'big-footed'. The enlarged legs are obvious among the 20 or so surviving species and are central to what makes them unusual: they dig burrows and construct mounds. Alone among birds, megapodes do not incubate their eggs by direct contact with the adult's body, but instead they use a number of alternative external sources to heat their eggs.

The best known species of megapodes are found in Australia, the Malleefowl (*Leipoa ocellata*), the Australian Brush-turkey (*Alectura lathami*) and the Orange-footed Scrubfowl (*Megapodius reinwardt*). However, the heartland of the family is in the Pacific Islands, where a diversity of forms occurs, some confined to single islands or island groups. There is good evidence that the current crop represents less than half the number of species that made it into modern times, and about 30 species have become extinct. Despite their ground-dwelling nature and gamebird-like build, megapodes are natural dispersers and are quite often seen flying across significant stretches of water to colonise offshore islands. The chicken lookalike didn't cross the road; it crossed the sea.

A classic example of the family is the Micronesian Scrubfowl (*Megapodius laperouse*) which occurs on Palau and in the northern Marianas islands, and was also once found on Guam. It is small, about 30cm long, rotund and mainly blackish in colour, with a paler greyish head and orange bill and legs. In common with most megapodes, birds on Palau build large mounds out of the soil, which act as incubators for the eggs. For such a modestly sized bird, the effort of building the mounds, just by kicking sand or earth with the feet, must be considerable, and their end product is frequently up to 7m in length, 6m in width and more than 1m high. There is, of course, more to the structure than just piles of material. In the majority of mounds the birds intermix a significant amount of organic matter, and it is the heat given off by micro-

organisms breaking down the plant material that actually provides the heat source to incubate the eggs. The function of the rest of the mound, usually sand in this case, is to keep the heat generated by the decomposition from leaking out of the mound, and also to provide a substrate in which the eggs can be concealed.

The act of building the mound is laborious, and so is egg-laying. Megapodes don't lay every day, but once every few days, sometimes with an interval of more than a week. Each time they have to dig a new hole in the mound, using their celebrated feet, to find a new spot for their latest addition to the clutch (the total number of eggs varies, but is certain to exceed ten at times). The whole process can take hours, especially if there is extra maintenance to perform, which means adding or removing substrate so that the temperature is reasonably constant inside – in the case of Micronesian

Above: The well-named Orange-footed Scrubfowl is part of the Megapode family, famous for their outsize feet.

Scrubfowl, it needs to be 30–33°C. It is likely that all megapodes are able to measure the temperature of the mound, although it is not known exactly how they do this. They often appear to 'taste' the mound materials first by dipping in the bill.

Intermittent modifications of the incubator are the limit of parental care in Micronesian Scrubfowl, leaving the adults to get on with their own priorities. Meanwhile, each egg is laid alone within the mound, and the hatchling similarly receives no support from parents or siblings. Fortunately, a young megapode hatches in a far more advanced stage of development than any other bird, a state sometimes dubbed as 'superprecocial'. Once hatched, young megapodes are able to cope with their world in a way that has no true equivalent among any other birds.

Opposite: They might not look agile, but Micronesian Megapodes fly well and can hop from island to island. Above: The Micronesian Scrubfowl may be the only bird in the world that practises three different types of incubation. Leaving the eggs in mounds, geothermally heated soil or in sand heated by the sun emancipates them from the bother of looking after young.

The hyper-developed state of young megapodes starts in the egg. Each egg is two or three times larger than the eggs of a bird of similar body size, and also contains far more yolk (more than 50 per cent of the egg volume) than equivalent species. The shell is thinner than for other birds, and has specially adapted pores that aid gaseous exchange in the egg's unusual buried state. The result that a megapode at hatching is something of a super-chick. Rather than politely chipping out of its egg with its egg-tooth, it instead kicks out with its feet. It climbs its way up through the sand and litter, apparently sometimes resting on its back and kicking upwards, at the same time allowing its down to dry. Once at the surface it is so well developed that it can actually fly almost immediately, which is unique among birds. In practice, however, a megapode chick is usually exhausted when it first reaches the surface and is very vulnerable. It is literally unsteady on its feet, however large they might be, for several hours.

It seems hard to believe that, during the heroic hatching of its offspring, an adult megapode will be out of contact, perhaps far away, and completely unaware of any success it has achieved as a breeding bird. That, though, is what mound-building does. It emancipates a megapode from the hard work

Above: The Orange-footed Scrubfowl is one of three Megapode species in Australia. It constructs mounds and uses the heat from microbial breakdown of vegetation to heat the eggs.

and risk of direct contact with eggs and chicks, but also emancipates it from experiencing the fruit of its labours.

The life-history of megapodes is unusual enough, but for the Micronesian Scrubfowl, there is a further short chapter to tell which, in some ways, marks out this species as still more unusual even than its peers. While we might see mound-building as an extreme adaptation to non-contact incubation, it isn't, in fact, the only one. The Micronesian Scrubfowl is a mound-builder on Palau, but on the Marianas Island group it practises something completely different.

The islands of the Pacific Rim are part of the Ring of Fire, a region of the world with heightened surface volcanic activity. On some islands there are geothermal vents, around which the ground is warmed to a consistent temperature. Although these are naturally limited in scale and space, they do afford opportunities for Micronesian Scrubfowl (and other species) to lay their eggs; in these cases the parent doesn't even need to tend a mound. All a female has to do is dig a pit, lay her egg and allow the warmth of the ground to do the rest. Elsewhere, some megapodes do something similar, but rather than selecting ground warmed by geothermal activity, they simply lay eggs in the sand and allow the sun to keep them warm. In at least some areas, it seems that the ever-adaptable Micronesian Scrubfowl does this too. If so, it is perhaps the only bird in the world with three separate strategies for incubating its eggs.

But how might mound-building have arisen in the first place, bearing in mind that it involves a profound behavioural switch that no other species of bird has followed? The theory goes that it probably developed from a habit of leaving the eggs temporarily covered by vegetation, for safety reasons, when an adult left the nest – as happens with grebes, for instance. From there the ancestral parents would have begun to spend less and less time with their eggs, as the knack of using heat from decomposition caught on. Finally, ancient megapodes abandoned their chicks completely.

Such a change could never, perhaps, have happened on continental land-masses, where such upstart adaptations would probably have fallen foul of predation – this would have been a time when all kinds of large animals, including birds, would have nipped the development in the bud. On oceanic islands, however, with no land predators, evolution often favours unusual branches of adaptation. It seems that 'incubation-lite' was one of these.

NEW CALEDONIAN CROW

The world's cleverest bird?

Does the world's most intelligent bird come from a medium-sized island in the south-west Pacific? It certainly seems that way. The New Caledonian Crow (*Corvus moneduloides*) has been described as the most advanced user of tools in the animal kingdom, aside from people, and tool-using is considered to be a measure of intelligence. If that's true it is certainly remarkable. The New Caledonian version of a crow doesn't look any different from most other species of crow, being essentially black all over. And in view of the fact that it does have close relatives all over the world, why should this bird, confined to its namesake island, be any more brilliant than the rest?

What we do know is that the New Caledonian Crow uses tools extensively, both in the wild and in captivity. In New Caledonia one of its favourite food items is the larvae of a wood-boring longhorn beetle in the family Cerambycidae, a species also only found on the island. These larvae are so energy-rich that it only takes a few of them to satisfy an individual crow's requirements for the day. They are found in cracks and holes in wood, and are difficult, or impossible, to extract using simply the bill. What the crows do, therefore, is to chivvy or drag the larvae out using sticks. A few other birds around do something similar, such as the Woodpecker Finch (*Camarhynchus pallidus*) of the Galapagos.

What makes the tool-use of the New Caledonian Crow unusual is that it goes one stage further and fashions its tools from diverse materials, each suitable for the occasion. The Woodpecker Finch usually just picks up a suitable-looking stick, which it uses repeatedly and may actually cut down to an ideal size – impressive enough. The New Caledonian Crow, though, is capable of creating a diverse tool-shed, with some tools made for specific tasks. And it can fashion tools out a wide range of different plant debris, including leaves, grasses, bamboo stems, twigs, thorny vines and fern stolons.

Those studying the birds have discovered that the tools come in three main categories: a 'normal' stick, similar to that used by the Woodpecker Finch; a stick with a hook at the end; and tools cut to incorporate the barbed edges of leaves from Screw Pines (*Pandanus* spp.). These tools may sound

***Opposite:** The New Caledonian Crow is the most sophisticated tool-user among all birds – and most other animals. This one is using a stick to extract invertebrate food.*

simple, but making them can take considerable time and practice – the birds, after all, only have their bills to work with. If a bird needs to make a tool with a single hook at the end, for example, it might have to manufacture its own hook by levering up part of the wood. Alternatively, if it is using a vine stem with many hooks, it will then need to remove those that it doesn't want. The final product tends to be standardised, depending on the material – another highly unusual aspect of this bird's tool-using.

Tools made from *Pandanus* leaves are even more closely fine-tuned, and investigators have discovered that these, also, come in three different types:

wide tools, narrow tools and stepped tools, which are wide at one end and narrow at the other. The leaves of *Pandanus* are long with barbed edges, and the crows simply pull off a strip along the edge, which is evenly broad or narrow; the barbs act as hooks to drag the prey out. The stepped, tapered tools are particularly fine craftsmanship. Once the crow has pulled off a strip, it snips and tears stepwise along the edge of the strip until one end is very fine and the other end is wider and easily held.

Exhaustive research has shown that the different types of tools made from *Pandanus* leaves have a distinct geographic distribution. At the same time there are no ecological, climate or other known correlates with the type of tool used, so the conclusion is that each population has its culture. Thus, birds in one part of the island may use stepped tools, while next door they use wide tools, and so on. At the same time, the tools themselves are standardised even between populations, so that the wide, narrow or stepped tools of one part of the island are the same as those on another. This hints that, over time, the tools may have gone through improvements at various stages, transmitted throughout populations. If so, the New Caledonian Crow would be the first non-human creature that shows culturally transferred modifications to tools.

The suggestion that that tool-making techniques (including improvements) are passed between generations is given credence by an unusual aspect of New Caledonian Crow biology. The young enjoy an exceptionally extended period of parental care. Recent studies have found that parents will regularly feed their young ten months after fledging, which is far more than other birds of similar size (in Carrion Crows (*Corvus corone*) from Europe this may last a month). Honing the fine skills needed for tool use needs a long apprenticeship.

Skills and intelligence are impressively evident in the wild, but it is perhaps in the laboratory that the exceptional capabilities of these birds have been most vividly demonstrated. The most famous example concerned two captive crows known as Betty and Abel. During one experiment, they were given a choice between a straight wire and a hooked wire in order to obtain some food, in this case some meat placed in a bucket in a vertical glass tube. Abel took the hooked wire, leaving Betty with the straight wire. Although she had never encountered wire before, Betty lodged one end of the wire and bent the other end into a hook without any kind of practice, and was able to extract the food by lifting the handle of the bucket. This almost immediate clever use of entirely novel material was a powerful example of the capabilities of this bird.

Further experiments have merely continued to demonstrate the New Caledonian Crow's genius. When faced with floating food on the surface of a glass tube, the birds, having previously been taught to drop stones into the water, learnt how to drop enough stones to raise the water level so that the prey was in reach – and, faced with large stones and small stones, always preferentially used the larger ones, thus accomplishing the task more quickly. Other individuals have shown they are able to use a mirror to locate food behind them, making the link between a reflection and the real world, which is almost unique among non-human animals. Still others have demonstrated the facility for 'meta-tool' use, in one case using a short stick to open a matchbox containing a longer stick, which was then used to obtain food. The ability to obtain food in two stages, the first of which did not obtain immediate reward, is unique to these birds and higher primates.

Experiments on these engaging characters continue to uncover the workings of their minds. A recent experiment has shown that the crows are able to attribute invisible causes to effects that they observe. In a series of experiments, the visible effect was a large stick being inserted through a gap into their cage, close to their feeding tray; the cause they couldn't see was the human being moving the stick. If the stick simply appeared from nowhere, they would react nervously and be wary of the feeding tray. If, however, they saw a human enter the hide next to them, the stick move and the human then leave the hide, they were able to attribute the appearing of the stick to a human being, despite not actually seeing the person manipulate the stick, and they were then happy to approach the feeding tray, unafraid.

Quite how far the intelligence and cognitive ability of the New Caledonian Crow stretches is still the subject of numerous experiments, and there is little doubt that we will have cause to marvel at their inventiveness for many years into the future.

But the central question remains: why on earth should this particular species of crow, on this one island, possess an intelligence apparently far beyond any other species in the world. That's a question that not even the most brilliant of humans can yet solve.

BLUE BIRD-OF-PARADISE

Figs and fruits turn paradise upside-down

'Birds of paradise' is a daringly evocative name to give to a family of
42 large songbirds related to crows. It's the kind of name that it must be
hard to live up to, being something you might see when you get to Paradise,
perfect beyond imagination. The members do their best; they are among the
most extravagantly plumaged of all birds on this side of heaven, exhibiting
a wider and more extravagant range of feather shapes and colours than any
other. But in fact their name is more to do with a quirk of earthly taxidermy
than celestial presence. When the first specimens of birds of paradise were
brought to Europe in the 16th century they had been heavily doctored for the
feather trade, and had neither legs nor innards. The belief arose that, since
they did not require food or a perch, these ethereal birds must float in the
airs of 'paradise'.

There are few birders who would not sell their souls for a glimpse of
one of them. Part of their allure is that many species are confined to New
Guinea, a difficult and expensive tropical treasure-trove to reach. But mostly
it is because they are almost peerless for those who like their birds vividly
coloured, wackily attired and given to outlandish displays (the cotingas of
the Neotropics are the only family that comes close). And that's without their
sounds, which are often loud, strange and, if the truth be told, not necessarily
fitting in well with a description of the heavenly realm.

If there is one species that is perhaps the epitome of this most opulent of
bird families, it is the Blue Bird-of-paradise (*Paradisaea rudolphi*). About the
size of a small crow, this bird has the enigmatic good looks typical of a film
star, as opposed to all-out, exaggerated beauty, at least when it goes about
its normal daily life. Both sexes have a jet-black body offset by metallic blue
wings and tail, somewhat brighter in the male. The head is boldly coloured,
with a powerful powder-blue bill and very prominent white crescents above
and below the eye. The adult male is only strikingly different to the female
when seen from behind, where he has a bundle of filamentous flank-plumes
below the tail, making him look as though it is wearing a suit over a tutu.
These plumes are an amber colour above and various iridescent colours,
from cobalt blue to violet, below. In addition to this gear, the two central

*Opposite: The flank feathers of the male Blue Bird-of-paradise
stick out beneath the 'normal' plumage.*

tail feathers are greatly elongated and would double the length of the bird if measured from bill to tail-tip. The very tips of these ribbons are spatulate and iridescent blue, sometimes glowing like tail-lights.

The point of all this bling is, of course, to impress a female, but getting one to the point of admiring any male's wares is a vexed process. First, an individual must acquire an exclusive territory, which he can only gain through experience and rising to the top of the tree through intense competition. Secondly, he must sing from a high perch to advertise his presence and availability within the territory – the song is a loud series of up-slurred bell-like sounds. And thirdly, he must set up his own display arena and perform in the hope of receiving interested visitors. This he may do for hours a day and for months on end, with only fitful visits from admirers. The site he chooses is almost always low down in the forest, just 1–3m above ground and with plenty of undergrowth vegetation above it. The owner will invariably 'garden' the site by stripping leaves away from the immediate vicinity, which tidies up the stage and embellishes the performance.

To us, the display of the Blue Bird-of-paradise is an absolute marvel, even if is scrutinised by those that matter with the same dispassionate intensity as shown by human judges at an Olympic gymnastics competition (and often

with the same baffling outcomes). It is one of the very few bird displays in which the performer hangs upside down, like a bat. The bird is one moment perching on a low branch, or a vine or liana, the next it slips backwards and is holding on, with his long tail-ribbons sticking up and arching down, making half a heart.

It is now obvious that the upside-down position is what shows off the male's flank plumes to maximum effect. Now fluffed up, they make a shimmering curtain of a rough 'V'-shape which looks rather like the tip of a giant peacock feather. There is an upper hemline of perfect cobalt-blue just below the feet, and a broader violet band below this. Dominating the centre of the chest is a black mark, like the pupil of an eye, which can look round or oval, depending on the degree of fluffing, and is bordered by iridescent blue towards the throat. The male Bird-of-paradise doesn't just hang there, of course, looking like a decoration, but enriches the visual feast by vibrating the feathers rapidly, giving the colourful ruff a coaxing shimmer. And it also provides a soundtrack that, at its zenith when a female is in attendance, fits the picture rather well, a strange, very quick repeating series of weird buzzing sounds. It would be no exaggeration to say that the singer could have changed from a bird to a clockwork toy, with his vibrating body movements and

*Opposite: With their powerful feet and strong build, birds-of-paradise are well adapted to plucking arboreal fruit from awkward positions. **Above:** The Blue Bird-of-paradise is one of the very few species of any bird to display upside-down.*

sounds that closely resemble some strange motorised device. This is possibly the most peculiar bird display in the world.

To hear and see the extraordinary efforts of a Blue Bird-of-paradise in the presence of a visiting female is surely one of the top experiences any birder could enjoy. But it also poses an intriguing question. How can such a display evolve, in which a male more or less devotes his life to perfecting moves that matter only for the brief moment of successful courtship? How can a male afford to grow his sumptuous feathers? How did it reach the point where hanging upside-down was a feasible part of the display?

Nobody knows the full answers to these questions, but there is one widely acceptable theory for the development of outlandish displays among birds-of-paradise and others. It is thought that their diet of fruit (and here, especially figs) emancipates them from the humdrum need to find enough food to survive the day. In tropical forests, with their agreeable temperatures and abundant super-nutritious fruit, as well as insects for protein, life is a little bit easy and foraging needn't take up much time. In such a climate, it is perfectly possible for a female to carry out all breeding tasks, including feeding the young unaided (although the clutches in birds-of-paradise are very small, just one egg in the case of the Blue species). Where a male is freed from any parental duties, it can concentrate entirely on becoming the body beautiful, the mate of choice among the local females. His life, essentially, revolves around copulating with as many females as possible.

Where such possibilities exist, competition is stiff, and only the best males get noticed. One way to stand out is to define your own territory, keeping rivals off your patch so that you own the court in which you display. Still another way is to spend longer at your display ground, to sing better than the rest, become more colourful, or to create more impressive moves. In each case natural selection favours the extravagant, the time-consuming and the star quality.

And a diet of fruit is the key.

Opposite: The male Blue Bird-of-paradise begins his display by falling back from a perched position.

EXTINCTIONS

Islands: lands of lost birds

There's an eerie birding experience available to anyone at the touch of a button. Try Cornell Laboratory of Ornithology's Macaulay Library (macaulaylibrary.org), and look up *Psittirostra psittacea*. There are several recordings of this bird, known by its Hawaiian name of ʻŌʻu, including one of the song recorded at the Koaie Stream Cabin at the Alakaʻi Swamp, in the highlands of the island of Kauaʻi in 1964. It isn't a great recording, but the bird has a fine territorial tune, with pure notes, some slightly slurred, and accomplished trills. One writer eulogises: 'In purity, sweetness and power the song of the ʻŌʻu far surpasses the canary's best efforts.' H.W. Henshaw, obviously a fan, goes on to muse: 'Unfortunately the ʻŌʻu, as a rule, is not very generous with its song, and too often the listener has to be content with snatches of melody in place of the finished performance.'

Oh, to be content with even a snatch of melody. That is no longer possible in the wild, or anywhere. After long years of population decline, the last confirmed living ʻŌʻu was seen on 17 February 1989 at the same location as the recording. It is one of the latest to join the grim list of birds that have been wiped out from Hawaii, including perhaps half of the 40 known Hawaiian honeycreepers. There are a host of reasons: deforestation, habitat degradation, introduced animals and plants, the introduction of mosquitoes causing avian malaria; cyclones and deadly volcanic floes. Wild Hawaii has, for many years, been in terminal decline. This mid-Pacific island chain started with a rich endemic flora and fauna, but one by one its treasures have diminished. And what makes it different to other places in the world is that the extinctions have become relentless, drip-feeding into the very recent past.

When we think of extinction, we sometimes forget what goes with it. That's partly because the most famous extinctions tended to happen long ago, from dinosaurs to the Dodo (*Raphus cucullatus*) to the Passenger Pigeon (*Ectopistes migratorius*), the last-named gone in 1914. We are speculative about the lives of ancient creatures, but some modern extinctions come with a lot more baggage. The birds that have died out are better known. They have personalities: behavioural, ecological, vocal – something to put on their gravestone. As living beings they had an effect on other parts of their world.

Take the ʻŌʻu once again. This was a chunky member of its family, 17cm long, with a powerful bill with a severe hook at the tip. It was dark olive-green in colour, and the male had a buttery-yellow head. It once occurred on every main Hawaiian island where its main food-plant was the abundant 'ieʻie tree

(*Freycinetia arborea*). The bird would eat its small fruits, its flowers and the leaf-bracts around the inflorescences. It would spend most of its time around these trees and was probably the most important pollinator of the 'ie'ie. 'O'us would also wander out of season and take the fruit of other native plants and, if the opportunity arose, it would feed on caterpillars when these had peaks in abundance. The 'O'u was not, though, a slave to native tastes; it would also take fruit from bananas, peaches and guava.

The point, though, is that the 'O'u evolved with native plants and undoubtedly played an important role in pollination and, perhaps, dispersal. Now that role is played by different, introduced species, or perhaps no

Above: The 'O'u is just one of the more recent on the conveyor belt of extinctions of Hawaiian landbirds. It was last seen in 1989.

species at all. Other Hawaiian honeycreepers fitted into other niches that are perhaps now vacant. The colourful 'Ula-'ai-hawane (*Ciridops anna*) fed on the inflorescences of a fan-palm known as the loulu palm (*Pritchardia* sp.) or 'hawane' on the island of Hawaii, while the Lana'i Hookbill (*Dysmorodrepanis munroi*) may have fed on the fruits of Opuhe (*Urera kaalai*) on its namesake island. A number of extinct species subsisted on nectar, including the Black Mamo (*Drepanis funerea*) of Molokai and the Mamo (*D. pacifica*) of the big island, both of which specialised on the abundant native lobelias, as well as the Lesser 'Akialoa (*Hemignathus obscura*). Other species, also now gone, fed on seeds, such as the Kona Grosbeak (*Chloridops kona*) of Hawaii. R.C.L.

Above: *The gorgeous I'iwi is a living reminder of the wondrous diversity of native Hawai'ian birds.*

●

Perkins, writing in 1895, said of this species feeding in Sandalwood trees (*Myoporum sandwicense)*: 'Its food consists of the seeds of the fruit of the aaka [Sandalwood]... and as these are very minute, its whole time seems to be taken up cracking the extremely hard shells of this fruit...The incessant cracking of the fruits when one of these birds is feeding, the noise of which can be heard from a considerable distance, renders the birds much easier to get than it otherwise would be'. The effect that the Kona Grosbeak had on its food-plants is not recorded. Neither do we have any idea how much of a role species such as the mamos had on the lobelias that they pollinated. There are still similar pollinators, such as the reasonably common 'I'iwi (*Drepanis coccinea*), around in modern-day Hawaii, and it would have been fascinating to know how each pollinator competed with the rest. This opportunity has passed us by.

If you go into a Hawaiian forest today, the question of what you don't see and experience resonates as much as what you cannot. Hawaii is noted for having no particularly pronounced dawn chorus – in season, song is consistent throughout the day rather than having a peak at first light. Could the dawn chorus have been wiped out along with the missing birds? Another intriguing dimension is mixed feeding flocks, such a feature of tropical forests worldwide. On the big island of Hawaii itself, there is a very short flocking season that occurs just after the breeding season – did flocks, especially of nectarivorous birds, once dominate the forests all year round? Certainly the island of Kaua'i was once known for its tightly co-ordinated, mixed flocks. These used to include such extinct forms as the Kaua'i Nukupu'u (*Hemignathus lucidus)* and the Kaua'i Akialoa (*H. stejnegeri*), brilliant green and yellow. Nowadays the flocks here are depleted and far more haphazard. The decline in the populations of all the components has undoubtedly played a part. And perhaps there is also no great necessity to form flocks? One of the functions of mixed feeding flocks is the collective need to detect predators – perhaps the potential predators have gone too?

One thing is for sure, the forests of Hawaii have had their heart ripped out. That is one aspect we forget about extinctions. The death of a species is the death of its ecology and song and everything about it. No bird is an island. When the species goes, its prey and companions and competitors change, and in the broad sense its habitat does too. When one personality goes, the personality of the forest changes too, bit by bit.

●

FURTHER READING

WEBSITES

In addition to the books below, there are numerous websites that provided invaluable information in the course of preparing this book. Among the most frequently used were:

Birds of North America Online *bna.birds.cornell.edu/bna*

Birdlife International *www.birdlife.org*

Bird Sounds *www.xeno-canto.org*

Science Daily *www.sciencedaily.com*

Google Scholar *www.scholar.google.co.uk*

BOOKS

Brooke, M. (2004). *Albatrosses and Petrels across the World*. Bird Families of the World 11. OUP, Oxford.

Cramp, S. (ed.) (1988) *The Birds of the Western Palearctic*. Vol 5. Tyrant Flycatchers to Thrushes. OUP, Oxford.

Cramp, S., Perrins, C. M. & Brooks, D.J. (eds). (1993).*The Birds of the Western Palearctic*. Vol 7. Flycatchers to Shrikes. OUP, Oxford.

Cramp, S., Perrins, C. M. & Brooks, D.J. (eds). (1994). *The Birds of the Western Palearctic*. Vol 8. Crows to Finches. OUP, Oxford.

Cramp, S. & Simmons, K.E.L. (eds.) (1983) *The Birds of the Western Palearctic*. Vol 3. Waders to Gulls. OUP, Oxford.

Davies, N.B. (2010). *Cuckoos, Cowbirds and Other Cheats*. A&C Black, London.

Davies, S.J.J.F. (2002). *Ratites and Tinamous*. Bird Families of the World 9. OUP, Oxford.

Davis, L.S. & Renner, M. (2003). *Penguins*. T. & A.D. Poyser, London.

del Hoyo, J., Elliott, A. & Christie, D.A. eds. (2003). *Handbook of the Birds of the World*. Vol 8. Broadbills to Tapaculos. Lynx Edicions, Barcelona.

del Hoyo, J., Elliott, A. & Christie, D.A. eds. (2007). *Handbook of the Birds of the World*. Vol 12. Picathartes to Tits and Chickadees. Lynx Edicions, Barcelona.

del Hoyo, J., Elliott, A. & Christie, D.A. eds. (2009). *Handbook of the Birds of the World*. Vol 14. Bush-shrikes to Old World Sparrows. Lynx Edicions, Barcelona.

del Hoyo, J., Elliott, A. & Christie, D.A. eds. (2010). *Handbook of the Birds of the World*. Vol 15. Weavers to New World Warblers. Lynx Edicions, Barcelona

del Hoyo, J., Elliott, A. & Christie, D.A. eds. (2011). *Handbook of the Birds of the World*. Vol 16. Tanagers to New World Blackbirds. Lynx Edicions, Barcelona.

del Hoyo, J., Elliott, A. & Sargatal, J. eds. (1994). *Handbook of the Birds of the World*. Vol 2. New World Vultures to Guineafowl. Lynx Edicions, Barcelona.

del Hoyo, J., Elliott, A. & Sargatal, J. eds. (1996). *Handbook of the Birds of the World*. Vol 3. Hoatzin to Auks. Lynx Edicions, Barcelona.

del Hoyo, J., Elliott, A. & Sargatal, J. eds. (1999). *Handbook of the Birds of the World*. Vol 5. Barn-owls to Hummingbirds. Lynx Edicions, Barcelona.

del Hoyo, J., Elliott, A. & Sargatal, J. eds. (2002). *Handbook of the Birds of the World*. Vol 7. Jacamars to Woodpeckers. Lynx Edicions, Barcelona.

Elphick, C., Dunning, J.B. Jr. & Sibley, D. (2001). *The Sibley Guide to Bird Life and Behaviour*. Chanticleer Press, Inc.

Ferguson-Lees, J. & Christie, D.A. (2001). *Raptors of the World*. Christopher Helm, London.

Frith, C.B. & Beehler, B.M. (1998) *The Birds of Paradise*. Bird Families of the World 6. OUP, Oxford.

Frith, C.B. & Frith, D.W. (2004). *The Bowerbirds*. Bird Families of the World 10. OUP, Oxford.

Gaston, A.J. (2004) *Seabirds: A Natural History*. T. & A.D. Poyser, London.

Hansell, M. (2000). *Bird Nests and Construction Behaviour*. CUP, Cambridge.

Hume, J.P. & Walters, M. (2012) *Extinct Birds*. T. & A.D. Poyser, London.

Kaufman, K. (1996). *Lives of North American Birds*. Peterson Natural History Companions. Houghton Mifflin, Boston.

Kirwan, G. & Green, G. (2011). *Cotingas and Manakins*. Helm Identification Guides. Christopher Helm, London.

Loon, R. & Loon, H. (2005). *Birds: The Inside Story. Exploring birds and their behaviour in southern Africa*. Struik, Cape Town.

Marchant, S. & Higgins, P.J. (co-ordinators). (1990). *Handbook of Australian, New Zealand and Antarctic Birds*. Volume 1. Ratites to Ducks. RAOU/OUP Australia.

Newton, I. (2008). *The Migration Ecology of Birds*. Elsevier/Academic Press, London.

Otter, K.A. ed. (2007) *Ecology and Behaviour of Chickadees and Titmice, An Integrated Approach*. OUP, Oxford.

Perrins, C.M. ed. 2003. *The New Dictionary of Birds*. OUP, Oxford.

Pratt, H. D. (2005) *The Hawaiian Honeycreepers*. Bird Families of the World 13. OUP, Oxford.

Rowley, I. & Russell, E. (1997). *Fairy-Wrens and Grasswrens*. Bird Families of the World 4. OUP, Oxford.

Simpson, K. & Day, N. (1999). *Field Guide to the Birds of Australia*. 6th edition. Penguin Books Australia, Ringwood, Vic.

Scott, G. (2010) *Essential Ornithology*. OUP, Oxford.

Tickell, W.L.N. (2000) *Albatrosses*. Pica Press, Sussex.

Wells, D.R. (1999). *The Birds of the Thai-Malay Peninsula*: Vol 1: Non-Passerines. Academic Press, London.

INDEX

A

Aggression
    Northern Wren 16-19
Alarm calls 70-71
Albatrosses 172-177
Andean Cock-of-the-rock (*Rupicola peruviana*) 138-143
Andean Hillstar (*Oreotrochilus estella*) 163
Ant-following 150-153
Antbirds 150-153
Arabian Babbler (*Turdoides squamiceps*) 80-83
Arctic Tern (*Sterna paradisaea*) 176
Ashy-headed Laughingthrush (*Garrulax cinereifrons*) 68
Atiu Swiftlet, (*Aerodramus sawtelli*) 87
Australian Brush-turkey (*Alectura lathami*) 200
Australian Swiftlet (*Aerodramus terraereginae*) 87

B

Bay-headed Tanager (*Tangara gyrola*)
Bee Hummingbird (*Mellisuga helenae*) 46
Beryl-spangled Tanager (*Tangara nigroviridis*) 154, 157
Bill specialisation
    Common Crossbill 32-35
    Eurasian Oystercatcher 36-39
    toucans 144-149
Bird larders 26-31
    and breeding success 31
Birds and people
    Cliff Swallow 124-127
    Southern Cassowary 98-103
Black Mamo (*Drepanis funerea*) 218
Black-backed Jackal (*Canis mesomelas*) 46
Black-capped Chickadee (*Poecile atricapillus*) 120-123
Black-faced Sheathbill (*Chionis minor*) 184
Black-nest Swiftlet (*Aerodramus maximus*) 84-85, 87
Black-tailed Jackrabbit (*Lepus californicus*) 130
Black-throated Mango (*Anthracothorax nigricollis*) 160, 161
Blue Bird-of-paradise (*Paradisaea rudolphi*) 210-215

Blue-and-black Tanager (*Tangara vassorii*) 155, 157
Blue-mantled Thornbill (*Chalcostigma stanleyi*) 163
Blue-necked Tanager (*Tangara cyanicollis*) 157
Boubous 56-59
Bowerbirds 98-103
Brood parasitism
    Great Spotted Cuckoo 20-25,
    Purple Grenadier 52-55
    Straw-tailed Whydah 52-55
Brood reduction, 166-171

C

Caciques (*Cacicus* spp.) 148
Cameroon Sunbird (*Cyanomitra oritis*) 45
Carrion Crow (*Corvus corone*) 208
Chestnut-eared Araçari (*Pteroglossus castanotis*) 148, 149
Christmas Island Glossy Swiftlet (*Collocalia esculenta natalis*) 84-85
Cliff Swallow (*Petrochelidon pyrrhonota*) 124-127
Commensal feeding
    Greater Racket-tailed Drongo 68-71
Common Crossbill (*Loxia curvirostra*) 32-35
Common Cuckoo (*Cuculus canorus*) 20
Common Poorwill (*Phalaenoptilus nuttallii*) 161
Communal foraging
    Harris's Hawk 128-131
Comparative ecology
    penguins 166-171, 178-183
    tanagers 154-157,
Conservation
    Hawaiian honeycreepers 216-219
Corn Bunting (*Emberiza calandra*) 25
Crimson-rumped Toucanet (*Aulacorhynchus haematopygus*) *147*
Crows
    host of Great Spotted Cuckoo 20-25

D

Dancing
    Wandering Albatross 188-193
Desert Cottontail (*Syvilagus auduboni*) 130

Display
    Blue Bird-of-paradise 210-215
    Great Bowerbird 98-103
Divorce
    Wandering Albatross 193
Dodo (*Raphus cucullatus*) 216
Duetting
    boubous/gonolek 56-59
Dusky Sunbird (*Cinnyris fuscus*) 44-45

E

Echolocation
    swiftlets 84-87
Edible-nest Swiftlet (*Aerodramus fuciphagus*) 87
Egg destruction
    Pheasant-tailed Jacana 79
Egyptian Vulture (*Neophron percnopterus*) 46
Emerald Tanager (*Tangara florida*) 155, 157
Emperor Penguin (*Aptenodytes forsteri*) 171, 178-183
Emu (*Dromaius novaehollandiae*) 104
Eurasian Oystercatcher (*Haematopus ostralegus*) 36-39
Evolution in action
    Cliff Swallow 124-127
    sunbirds 42-45,

F

Fairywrens 94-97
Falkland Skua (*Stercorarius antarcticus*) 166
Fan-tailed Widowbird (*Euplectes axillaris*) 64-65
Feet
    Micronesian Scrubfowl 200-205
    Southern Cassowary 104-107
Flight style
    albatrosses 172-177
    hummingbirds 158-163
Flocking
    Greater Racket-tailed Drongo 70
Flower gifts
    fairywrens 94-97
Food-storing
    Black-capped Chickadee 120-123
    Great Grey Shrike 26-31

Footed-ness
 Common Crossbill 32-35

G

Galapagos Penguin (*Spheniscus mendiculus*) 178-183
Gentoo Penguin (*Pygoscelis papua*) 166, 168
Giant petrels (*Macronectes* spp) 184
Glistening-green Tanager (*Chlorochrysa phoenicotis*)154
Golden Eagles (*Aquila chrysaetos*) 128
Golden-eared Tanager (*Tangara chrysotis*) *157*
Golden-naped Tanager (*Tangara ruficervix*) 157
Great Bowerbird (*Chlamydera nuchalis*) 98-103
Great Grey Shrike (*Lanius excubitor*) 26-31
Great Hornbill (*Buceros bicornis*) 70
Great Spotted Cuckoo (*Clamator glandarius*) 20-25
Greater Racket-tailed Drongo (*Dicrurus paradisaeus*) 68-71
Green-and-gold Tanager (*Tangara schrankii*) 155, 157
Group-living
 Arabian Babbler 80-83
 White-winged Chough 90-93,

H

Harris's Hawks (*Parabuteo unicinctus*) 128-131
Hawaiian honeycreepers 216-219
Hierarchy
 Varied Sitella 108-111
Honeyeaters 42
Hooded Crow (*Corvus cornix*) 22
Huddling
 Northern Wren 16-19
 Varied Sitella 110-111
Hummingbirds 42, 84, 158-163
'I'iwi (*Drepanis coccinea*) 218, 219
Impaling prey
 Great Grey Shrike 26-31
Incubation
 Micronesian Scrubfowl 200-205
 Ostrich 46-51
 Pheasant-tailed Jacana 76-79
 Rockhopper Penguin 166-171
Infanticide
 Pheasant-tailed Jacana 76-79

Intelligence
 New Caledonian Crow 206-209

J, K

Jungle Babbler (*Turdoides striata*) 68
Kaua'i Akialoa (*Hemignathus stejnegeri*) 219
Kaua'i Nukupu'u (*Hemignathus lucidus*) 219
Kelp Gulls (*Larus dominicanus*) 184
Kidnap
 White-winged Chough 93
King Penguins (*Aptenodytes patagonicus*) 166, 171
Kona Grosbeak (*Chloridops kona*) 218

L

Lana'i Hookbill (*Dysmorodrepanis munroi*) 218
Lanner Falcon (*Falco biarmicus*) 80
Lek: Andean Cock-of-the-rock 138-143
Lesser 'Akialoa (*Hemignathus obscura*) 218
Lettered Araçari (*Pteroglossus inscriptus*) 145
Light-mantled Sooty Albatross (*Phoebetria palpebrata*) 174, 177
Long-tailed Widowbird (*Euplectes progne*) 60-64

M

Magellanic Penguins (*Spheniscus magellanicus*) 166, 168
Magpie-lark (*Grallina cyanoleuca*) 59
Magpie
 host of Great Spotted Cuckoo 20-25
Malachite Sunbird (*Nectarinia famosa*) 44-45
Malleefowl (*Leipoa ocellata*) 200
Mamo (*Drepanis. pacifica*) 219
Marbled Murrelet (*Brachyramphus marmoratus*) 132-135
Mass roosting
 Northern Wren 17-19
Megapodes 200-205
Meliphagidae see Honeyeaters
Memory
 Black-capped Chickadee 120-123
Micronesian Scrubfowl (*Megapodius laperouse*) 200-205
Migration
 albatrosses 172-177

Yellow-browed Warbler 72-75,
Mimicry
 Great Spotted Cuckoo 24
Mobbing
 Greater Racket-tailed Drongo 68-71
Mountain Swiftlet (*Aerodramus hirundinaceus*) 87

N

Nest
 Ostrich 46-51
 Pheasant-tailed Jacana 76-79
 White-winged Chough 91-93
Nest-robbing
 toucans 144-149
Nest-site
 Cliff Swallows 124-127
 Marbled Murrelet 132-135
New Caledonian Crow (*Corvus moneduloides*) 206-209
Nocturnal lifestyle
 Swallow-tailed Gull 196-199
Northern Double-collared Sunbird (*Cinnyris reichenowi*) 43, 45
Northern Royal Albatross (*Diomedea epomophora*) 176
Northern Wren (*Troglodytes troglodytes*)16-19

O

Ocellated Antbird (*Phaenostictus mcleannani*) 150, 151
Oilbird (*Steatornis caripensis*) 84
Opal-rumped Tanager (*Tangara velia*) 157
Optimal foraging
 Swallow-tailed Gull 196-199
Opuhe (*Urera kaalai*) 218
Orange-billed Babbler (*Turdoides rufescens*) 68
Orange-footed Scrubfowl (*Megapodius reinwardt*) 200, 201, 204
Oropendolas (*Psarocolius* spp.) 148
Ostrich (*Struthio camelus*) 46-51, 104
O'u (*Psittirostra psittacea*) *216-217*

P

Pair-bonds
 albatrosses 172-177, 188-193
 fairywrens 94-97
 Great Grey Shrike 26-31
Papuan Swiftlet (*Aerodramus papuensis*) 87

Parasitism
    Straw-tailed Whydah 53-54
    Great Spotted Cuckoo, 20-25
Parental care
    Pheasant-tailed Jacana 76-79
    Rockhopper Penguin 166-171
    Marbled Murrelet 132-135
Passenger Pigeon (*Ectopistes migratorius*) 216
Pheasant-tailed Jacana (*Hydrophasianus chirurgus*) 76-79
Plain-pouched Hornbill (*Aceros subruficollis*) 71
Polymorphism
    White-throated Sparrow 116-119
Potato Bush (*Solanum ellipticum*)
    Spotted Bowerbird 102
Purple Grenadier (*Uraeginthus ianthinogaster*) 52-55
Purple-back Flying Squid (*Sthenoteuthis oualaniensis*) 199

R
Rainbow-billed Toucan (*Ramphastos sulfuratus*) 144-145
Red-eyed Vireo (*Vireo olivaceus*) 56, 59
Resource partition
    Oystercatcher 36-39
    Tanagers 154-157
Road casualties
    Cliff Swallow 125-126
Roadrunner (*Geococcyx californianus*) 20
Roosting
    hummingbirds 158-163
    Northern Wren 16-19
    Varied Sittella 110-113
Ruddy Turnstone (*Arenaria interpres*) 108
Rufous-capped Ant-thrush (*Formicarius colma*) 153

S
Scale-backed Antbird (*Hylophylax poecilonotus*) 153
Scavenging
    Sheathbills 184-187
Sentinel duty
    Arabian Babblers 80-81
Seven-coloured Tanager (*Tangara fastuosa*) 154
Sexual selection
    Long-tailed Widowbird 60-65
Sheathbills 184-187

Shy Albatross (*Thalassarche cauta*) 173, 177
Skuas (*Stercorarius* spp.) 184
Slate-coloured Boubou (*Laniarius funebris*) 56, 58-59
Snowy Sheathbill (*Chionis albus*) 166
Song mimicry
    Straw-tailed Whydah 52, 54
Southern Cassowary (*Casuarius casuarius*) 104-109
Southern Rockhopper Penguin (*Eudyptes chrysocome*) 166-171
Southern Royal Albatross (*Diomedea sanfordi*) 176
Sparkling Violetear (*Colibri coruscans*) 158,159
Splendid Fairywren (*Malurus splendens*) 94-97
Spot-winged Antbird (*Schistocichla leucostigma*) 153
Spotted Antbird (*Hylophylax naevioides*) 152, 153
Spotted Bowerbird (*Chlamydera maculata*) 102
Spotted Hyaenas (*Crocuta crocuta)* 46
Spotted Tanager (*Tangara punctata*) 157
Status
    Arabian Babbler 83
Storing food
    Black-capped Chickadees 121-123
Straw-tailed Whydah (*Vidua fischeri*) 52-55
Sunbirds (Nectariniidae) 42-45
Superb Fairywren (*Malurus cyaneus*) 94-97
Swallow-tailed Gull (*Creagrus furcatus*) 196-199
Swifts and swiftlets 84-87

T
Tail length
    Widowbirds 60-65
Tanagers 154-157
Tool use
    New Caledonian Crow 206-209
Toucans 144-149
Tree Tobacco (*Nicotiana glauca*)
    hummingbirds 43
    sunbirds 43-45
Tropical Boubou (*Laniarius major*) 56
Tufted Coquette (*Lophornis ornatus*) 160, 161

Turquoise Tanager (*Tangara mexicana*) 156

U, V
'Ula-'ai-hawane (*Ciridops anna*) 218

Vagrancy
    Yellow-browed Warbler 72-75
Varied Sitella (*Daphoenositta chrysoptera*) 108-111
Vocalisations
    boubous/gonolek 56-59
    White-throated Sparrow 116-119

W
Wandering Albatross (*Diomedea exulans*) 172, 173, 188-193
Western Bonelli's Warbler (*Phylloscopus bonelli*) 73
White-chinned Woodcreeper (*Dendrocincla merula*) 153
White-plumed Antbird (*Pithys albifrons*) 150
White-throated Antbird (*Gymnopithys salvini*) 153
White-throated Sparrow (*Zonotrichia albicollis*) 116-119
White-winged Chough (*Corcorax melanorhamphos*) 90-93
Widowbirds 60-65
Willow Tit (*Poecile montanus*) 123
Wing length
    Cliff Swallows 127
Woodpecker Finch (*Camarhynchus pallidus*) 206

Y
Yellow-bellied Tanager (*Tangara xanthogastra*) 157
Yellow-browed Warbler (*Phylloscopus inornatus*) 72-75
Yellow-crowned Gonolek (*Laniarius barbarous*) *56-57*
Yellow-eyed Penguin (*Megadyptes antipodes*) 171
Yellow-throated Toucan (*Ramphastos ambiguus*) 146, 147
Yellowhammer (*Emberiza citrinella*) 56

ACKNOWLEDGEMENTS

PHOTO CREDITS

The author and publishers are grateful to the following for permission to use their photographs.

1 David Tipling/FLPA; 2 Frans Lanting/FLPA; 5 Tim Laman/naturepl.com; 9 Frans Lanting/FLPA; 10 Richard Du Toit/Minden Pictures/FLPA; 12 Â© Biosphoto , Patrice Correia/Biosphoto/FLPA; 14 Dave Pressland/FLPA; 17 Richard Brooks/FLPA; 19 Markus Varesvuo/naturepl.com; 21 Roger Powell/naturepl.com; 22 Yoram Shpirer; 23 Jose B. Ruiz/naturepl.com; 25 Roger Powell/naturepl.com; 27 Wim Weenink/Minden Pictures/FLPA; 28 Duncan Usher/Minden Pictures/FLPA; 30 Wim Weenink/Minden Pictures/FLPA; 33 Photo Researchers/FLPA; 34 Adri Hoogendijk/Minden Pictures/FLPA; 37 Paul Hobson/FLPA; 40 Richard Du Toit/Minden Pictures/FLPA; 43 Neil Bowman/FLPA; 44 Neil Bowman/Shutterstock; 47 Martin Maritz/Shutterstock; 48 Frans Lanting/FLPA; 49 Chris & Tilde Stuart/FLPA; 51 Suzi Eszterhas/Minden Pictures/FLPA; 53t Nik Borrow; 53b Neil Bowman/FLPA; 57 Ignacio Yufera/FLPA; 58 Michael Gore/FLPA; 60, 62 Richard Du Toit/Minden Pictures/FLPA; 65 Neil Bowman/FLPA; 66 Martin Hale/FLPA; 69 Bernard Castelein/naturepl.com; 71 Chien Lee/Minden Pictures/FLPA; 73 Harri Taavetti/FLPA; 74 Markus Varesvuo/naturepl.com; 77 GoPause/Shutterstock; 78 Erica Olsen/FLPA; 81 Hanne & Jens Eriksen/NPL; 82 Miles Barton/naturepl.com; 85t Ingo Arndt/naturepl.com; 85b Phil Chapman/naturepl.com; 88 Konrad Wothe/naturepl.com; 91 William Osborn/naturepl.com; 92 Ian Montgomery; 95 Neil Bowman/FLPA; 96 Martin Willis/Minden Pictures/FLPA; 97 Rob Drummond, BIA/Minden Pictures/FLPA; 99 Ingo Arndt/Minden Pictures/FLPA; 100t David Hosking/FLPA; 100b Michael Gore/FLPA; 103 Konrad Wothe/Minden Pictures/FLPA; 105 Dave Watts/naturepl.com; 106t Kevin Schafer/Minden Pictures/FLPA; 106b Kerstiny/Shutterstock; 109 Dave Watts/naturepl.com; 111 Ian Montgomery; 112, 113 Peter Ware; 114 Frans Lanting/FLPA; 117 Scott Leslie/Minden Pictures/FLPA; 119 Jose Schell/naturepl.com; 121 Philippe Henry/Biosphoto/FLPA; 123 Elliotte Rusty Howard/Shutterstock; 125 S & D & K Maslowski/FLPA; 126 Frans Lanting/FLPA; 129 Tom Vezo/Minden Pictures/FLPA; 130 ImageBroker/Imagebroker/FLPA; 133 Konrad Wothe/Minden Pictures/FLPA; 135 Mark Moffett/Minden Pictures/FLPA; 136 Kevin Elsby/FLPA; 139 Tui De Roy/Minden Pictures/FLPA; 141 Murray Cooper/Minden Pictures/FLPA; 142 Murray Cooper/Minden Pictures/FLPA; 145 Eduardo Rivero/Shutterstock; 146 Mark Caunt/Shutterstock; 147 Murray Cooper/Minden Pictures/FLPA; 149 Rich Lindie/Shutterstock; 151, 152 Christian Ziegler/Minden Pictures/FLPA; 155, 156 Murray Cooper/Minden Pictures/FLPA; 159 Michael & Patricia Fogden/Minden Pictures/FLPA; 160t Robin Chittenden/FLPA; 160b Frans Lanting/FLPA; 163 ImageBroker/Imagebroker/FLPA; 164 Frans Lanting/FLPA; 167 Tui De Roy/Minden Pictures/FLPA; 169t Frans Lanting/FLPA; 169b Konrad Wothe/Minden Pictures/FLPA; 170 Pete Oxford/Minden Pictures/FLPA; 173 Frans Lanting/FLPA; 174 Martin Hale/FLPA; 175 Jaap Vink/Minden Pictures/FLPA; 177 Chris & Monique Fallows/naturepl.com; 179 Fritz Polking/FLPA; 180 Frans Lanting/FLPA; 182 Kevin Schafer/Minden Pictures/FLPA; 183 Pete Oxford/Minden Pictures/FLPA; 185, 186 David Hosking/FLPA; 189 Kevin Schafer/Minden Pictures/FLPA; 190t Yva Momatiuk & John Eastcott/Minden Pictures/FLPA; 190b, 193 Frans Lanting/FLPA; 194, 197, 199 Tui De Roy/Minden Pictures/FLPA; 201 Konrad Wothe/Minden Pictures/FLPA; 202 Tim Laman/naturepl.com; 203 Wikipedia Commons/Michael Lusk; 204 Martin Willis/Minden Pictures/FLPA; 207 Roland Seitre/naturepl.com; 211 Alain Compost/Biosphoto/FLPA; 212, 213 Tim Laman / National Geographic Stock/naturepl.com; 214 Tim Laman/naturepl.com; 217 Wikipedia Commons; 218 Frans Lanting/FLPA.

AUTHOR ACKNOWLEDGEMENTS

Writing a book might be a solitary task, but it cannot be done without support of all kinds. To this end I wish to convey my heartfelt thanks to my wife Carolyn and to children Emmie and Sam for loving me throughout and keeping me sane.

Talking of sanity, I wish to thank Lisa Thomas, whose patience I tested to the limit at times on this project. The fact that this book got finished is a testament to her professionalism and perseverance.

Finally, thanks to all those researchers and scientists who strive to find out more about birds, inadvertently providing material to writers such as me.